# OCD Today

# Obsessive Compulsive Disorder

Revision Copyright © 2012 by José Espaillat, M.D.
Copyright © 2006 by José Espaillat, M.D.

All rights reserved. No part of this book may be reproduced or utilized in any form or by any means electronic or mechanical, including photocopying, recording or by any information storage and retrieval system, without permission in writing from the author and publisher.

Library of Congress Cataloging-In-Publication Data

Espaillat Inchautegui, José Francisco, M.D.
    OCD Today
    Obsessive Compulsive Disorder

Txu001305942                        2006

ISBN
978-0-578-00761-8

Printed by lulu

# Dedication

I want to dedicate this book to Erica, who is not only the greatest wife a man can have but also the greatest person that one can be around. She has been a source of inspiration and support. She has also been my graphic artist and editor.

I am grateful to JoAnn Rockefeller for gracefully giving her time to edit and provide detailed recommendations for the improvement of this book.

I have to thank my ancestors, those who I have met and those who I have not. Knowing about the way they lived their lives serves as inspiration and moral support. It helps me believe in myself and propels me forward to meet my self set responsibilities.

I acknowledge and thank all the researchers that spend years working in laboratories with animals and people. They answer the very complicated short questions. Over time they slowly build the links of a chain of knowledge that helps us understand the world around us and within us. This book is based on their work.

# Index:

| | | |
|---|---|---|
| I- | Introduction | 1 |
| II- | Living with OCD | 7 |
| III- | Epidemiology | 25 |
| IV- | Obsessions | 28 |
| V- | Compulsions | 29 |
| VI- | Traditional Definition of OCD | 30 |
| VII- | What is OCD? A biological theoretical approach | 31 |
| VIII- | Procedural behaviors | 37 |
| IX- | Ethology | 44 |
| X- | Most common compulsions and obsessions | 47 |
| XI- | Neuroanatomy and physiology | 63 |
| XII- | Other medical disorders associated with OCD symptoms | 70 |
| | a. Postencephalitis Parkinsonian Syndrome | 72 |
| | b. Huntington's Chorea | 73 |
| | c. Bilateral Necrosis of the Globus Pallidus | 73 |
| | d. Tourette's Disorder | 74 |
| | e. Pediatric Autoinmune Neuropsychiatric Disorders Associated with Streptococcus | |

|  |  |  |
|---|---|---|
|  |  | (PANDAS)......74 |
|  | f. | Hemiballism......76 |
|  | g. | Orbito-Frontal Meningioma......76 |
|  | h. | Psychosurgery......77 |
| XIII- | Neuroimaging and other studies ......81 |  |
| XIV- | Pathophysiology......91 |  |
| XV- | Treatment ......108 |  |
| XVI- | Conclusion ......120 |  |
| XVII- | Future research ......123 |  |
| XVIII- | Bibliography......125 |  |

# Introduction

Psychiatry, as in all fields in science, changes over time. Phenomenological observations (the phenomenon, the observation of what actually happens as opposed to how or why it happens) were described first. Initially, a few individuals with a fascination for human behavior, intelligence, and keen observation capacities, embarked in a quest to understand the underpinnings of emotions, thoughts and behaviors. With limited scientific methodology and technological tools, they studied extremely complex processes. These students of human behavior made clever observations, many of which still stand true and relevant today. These individuals tried to explain what was happening inside the mind or brain, from their external observations of behaviors and the verbalization of thoughts. They also attempted to answer why. They developed numerous theories, some of which, as science has been advancing, have been proven misguided and needed adjustments and restructuring. With other theories and observations, terminology has become more refined but the essence has remained the same.

Not too many years ago doctors believed, and some people still do, that the development of Schizophrenia was causally related to the mother's behavior. Doctors believed that people developed Schizophrenia because of their relationship with family members, especially the mother. There is the very early observation in the mental health field that some family behaviors appear to have an effect in the symptoms of their loved one with the illness. Today it is known that some families' interactions can affect the course of illness in Schizophrenia by triggering more symptoms, but it is not the cause of the illness.

One of the early theories about Obsessive Compulsive Disorder (OCD) explains that this is a disorder caused by deep intra-psychic conflicts, secondary to the person's experiences, that later express themselves as OCD. This theory was created by very insightful and intuitive people who gave an explanation to their observations. The theory is enticing since it makes sense. Sounding logical in science doesn't always mean that the information is factual, and this is what modern psychiatry is discovering with regards to this theory of OCD.

OCD has been extensively studied over time and has developed from early observational theories to more scientific and evidence based theories. Today OCD is believed to be a set of genetically inherited motor and cognitive procedural strategies that contribute most commonly to hoarding, creating or maintaining order, checking for danger, and/or reducing contamination.

With current knowledge, scientists are more careful about reaching premature explanations for the development, course, and prognosis of the very complex and different mental illnesses. In science, theories change and evolve, and this is expected and welcomed. Theories will continue to emerge but they should be viewed as the best theories at the moment and not as unquestionable truths.

The purpose of this book is to help families, patients, and mental health professionals understand the current ideas about OCD. By achieving this, people will view OCD as a natural phenomenon, which can happen to anyone; although there could be predisposing factors. We hope to help destigmatize OCD, decrease the guilt and shame associated with it, and help advance its treatment and future research. Our expectation is that this might serve as a link in a long and slowly built chain of knowledge.

The development and control of behaviors and emotions is probably the most complex of the brain functions. There are subtypes of OCD which makes OCD even more difficult to study and understand. There are multiple factors affecting the development and course of this condition. These factors have been studied by many people from different perspectives, giving rise to several different theories that may appear to contradict each other, but probably are complementary to each other.

This book will describe the most recent neurobiological theory of OCD as well as the most effective and proven treatments. Over time some theories may be proven wrong, as this is science, and this will be welcomed and expected. Areas where further research may lead to improved treatments will also be discussed.

# Living with OCD

The following narratives give the reader an idea of how this condition can affect someone's life. The stories are real although the names have been changed to protect confidentiality.

### Pam's story

On one occasion I was asked to evaluate a woman who was hospitalized. When I arrived to the patient's bedside, I found Pam, an emaciated twenty-seven year old woman with black hair and contrasting paper white skin, staring up at me with anxious and pleading eyes. Her husband, who was also at her bedside, appeared concerned. She explained that she was in the hospital for a possible back surgery but that was not the reason she had asked to see a psychiatrist. They went on to explain that approximately six years ago she had started checking the locks of their home multiple times, before going to bed, to make sure they were locked. She was doing this several times consecutively. She locked the doors and on her way to bed felt the overwhelming need to go back and check if the doors were truly locked, although she was aware that she had just locked them. Pam explained that sometimes she went to bed and then got out of bed to check the locks again. If she tried to resist her urge or desire to recheck the locks, she became intolerably anxious until she returned to recheck them. This urge repeated multiple times. Within a couple of months this need to check the locks gradually expanded in time until it was lasting approximately two hours. After two hours she felt more relaxed, stopped checking the locks and eventually fell asleep. By then her husband was usually asleep as she tried not to disturb him.

This behavior started affecting Pam at work as she was tired during the day due to lack of sleep at night. After several months of continuing this night time ritual, she began to feel the same need to check the locks every time she left her home as well as before bedtime. It took her two hours in the morning and two hours in the evening before the urge subsided. This behavior continued for two years and then the urge became so severe that she had to check the locks every time the doors were opened. This took her approximately four hours in the morning and four hours in the evening before she felt relieved and relaxed.

As the lock checking compulsion escalated, she was forced to arise earlier in order to check the locks and still make it to work on time. In spite of her efforts, she was often late to work. She found herself living a severely stressful life. She felt ashamed as did her husband.

Prior to her hospitalization, she had sought treatment at a Community Mental Health Center. While being treated at this clinic she took the antidepressant Prozac (at a dose she couldn't remember), which temporarily decreased her symptoms for a few months. However, the symptoms returned even while she was taking her medication. At this point she stopped treatment because of a combination of factors. The stigma of going to the clinic was overwhelming, she was feeling that her situation was hopeless, and the time that she spent checking the locks every time she left the house and came back made it nearly impossible. By the time I met her, she had refused to leave the house for approximately four years. Her lock checking compulsion had extended to four hours each time her husband left the home, returned, and/or any time a door was opened. They both knew that checking the locks repeatedly did not make any sense, but she could not stop doing it. They were desperate! The lock checking compulsion described above is an expression of one of the four most common compulsions associated with OCD; being safe or checking for danger.

### Brian's story

Brian was a fifty year old man when I met him. He had been in the military, had a master's degree, and lived with his wife Erma and his twelve year old son, Peter. Brian had been married before and had Peter from that prior relationship. Erma was Brian's high school sweetheart and their love rekindled when they met again during a high school class reunion. The family came to see me for evaluation and treatment of Peter who was having several difficulties, mostly of behavioral and academic performance. During the course of evaluation and treatment it came to light that the marriage was severely strained.

Brian had accumulated multiple items throughout the years and was unable to part with them. Approximately five years before I met them, Brian had bought three hundred outdated computers with their large monitors. He bought them cheaply at a government auction. Brian bought them thinking that he could make a profit by selling them. Approximately five years later he had not sold a single computer. He had them in his living room, bedrooms, kitchen, closets, basement, and on the dining table. Some of them were in different stages of disassembling.

Brian had saved all of his military paperwork, including every instructional material he ever received. "In case my son ever joins the military, he could use them." He would save every newspaper and magazine. "In school they always ask children for pictures and news for projects, so they could be useful." When he was confronted with the fact that Peter could get these items if he needed them from the internet or the library, he agreed but seemed at a loss and still unable to throw his newspapers and magazines away. He saved every empty glass bottle. "They can be useful to store things."

The family could not use their garage to park their cars for Brian had an old engine in the way that had been there for years. Brian's brother also had a dilapidated car in Brian's garage that he planned to work on. They were able to get his brother to take this car away. Needless to say the house was severely cluttered. They could not find things when they needed them, and even Brian did not have a clear inventory of all the things he had accumulated. Erma married Brian thinking that he needed a female presence in the home and that she would quickly make the house neat with his help. She tried many times to get rid of some of the items but during the process Brian was always able to save most of the things stating that they could be useful.

This letter is a summary written by him about the situation, and they agreed to allow me to include it in my book.

*We have been married for five months and my wife doesn't adapt to her new stage of life. I am anxious, stressed out, with desires of ... life. Because she recriminates to me about the following:*

1) I have many computer monitors, mouse, keyboards, copiers, around the house. Most of them were taken to a little shack outside the house and others are in the basement.

2) I have many screws, plumbing, electricity, carpentry, and computer effects around the house.

3) I have boxes with thousands of double prints of pictures.

4) I have boxes with my college books. (He has saved them for over thirty years).

5) I have boxes filled with old magazines and newspapers, in case my son decides to use them.

6) I have papers and books about my military career in case my son decides to go into the Armed Forces.

7) I have boxes with receipts for water, electricity, and alimentary pension for my daughters (his daughters are now approximately thirty years old). I have not

had time to look and throw unnecessary things away.

8) I save the medicines bottles, and the literature about them. I also save cards and advertisements, since I love to read.

9) I save, in case they are needed, bottles of olives, juices, etc...

10) I have approximately five radios all around the house and they all work. I love music.

11) I have approximately six television sets. A broken one in each room.

12) In the living room I have tools to fix computers.

13) I have so much clothes that I have them in three or four different parts of the house.

14) I have many cameras that do not work but maybe they could be used for parts.

15) I have medications in my room, and the kitchen.

16) The dinning table is almost always filled with papers. Ah, and I have an office equipped with everything.

17) Every time I go to the Veterans' Administration building, I stay up until twelve to one in the morning preparing my files.

18) I have computer CD programs, videos, and movies in my office, bedrooms and bathroom.

19) I am always sick because I have ... And when I get sick I don't want to contaminate my wife. This is why when I had the flu I didn't have sex with her for ten days. Even though we were just married I abstained myself.

20) Several weeks ago I got a urinary tract infection. I told my wife that my prostate was sick, that I had cancer (initially that's what I thought I had). On Sunday I looked in the internet and I thought it was cystitis. She had to have a urine test done. Her urine labs were fine but since it said "few bacteria", I had

her go to the internal medicine doctor, even though she had already gone to her gynecologist. The good thing about all this is that I did the recommended treatment twice. I took a lot of care not to transmit the infection to her even though the doctor had told me that it was not contagious.

21) I don't know why she gets upset. She had nasal allergy and I didn't want to kiss her because of fear of getting sick. Well, now I'm the one that has a cold, and I swear doctor that I don't want her to get it.

22) I feel bad when she recriminates to me for the towels lying around getting air, the shoes in the staircase, kitchen and living room. I can't bend over.

23) I can't throw away the many broken bicycles that I have, a table and a broken bench. They could all have some use.

I don't know if I can take her any longer. Help her please. She has very little tolerance.

Erma was thinking about leaving the family because she could not accept their living condition and felt unable to change it.

Brian had traits of the compulsion of cleanliness with the obsession of contamination, but his main compulsion was hoarding. It is common for people to have more than one compulsion.

The narrative just described is an example of a moderate case of hoarding. I call it a moderate case as many people with OCD, hoarding subtype, often have to move out of their homes because they can no longer fit in them. They have so much clutter that they have to walk through narrow trails of boxes and other items. Their living situation is dangerous and unsanitary. They move to a new home and do the same thing all over again. Sometimes they come to the attention of psychiatrists only after social services or the police intervene.

## Edward's story

Edward was a fourteen year old adolescent. He was hospitalized in a psychiatric inpatient unit when I met him, after being stabilized in a pediatric inpatient unit. He was hospitalized because he was not able to eat and had lost a severe amount of weight.

Edward's compulsions started suddenly and were initially mild. In the beginning stages, Edward had to count some items in sets of threes. The compulsions progressively worsened to where he needed to count everything in sets of threes and this consumed his time. Edward counted tiles on the floor or the ceiling, door knobs, light switches, etc...

Edward later developed a compulsion to have to do things in a very particular way. When he walked, it had to be in a way that felt right to him. Because of this compulsion of having to walk in a particular way, he tried to take the first step and had to immediately move back to do it again. It took him a long time to take a few steps. To an observer, Edward's way of walking probably looked worse than if he was walking on a mine field. His walking looked very bizarre.

Contrastingly he would walk fairly normal if someone was pulling him along.

Edward later developed a compulsion consisting of chewing the food a certain number of times and in a specific way that had to feel right to him. It was very difficult for him to feel right about what he was doing and he was taking a long time to eat small amounts of food. Edward was not eating enough to meet his metabolic demands and this caused significant weight loss, prompting the hospitalization. These compulsions were accompanied with the feeling that if he did not take these measures something horrible was going to happen to the world. Like the universe's balance was going to be lost. He understood at an intellectual level that this did not make any sense but still the feeling was overwhelming and he could not overcome it.

Edward has a case of OCD with compulsion of symmetry. Everything had to be done in a specific way that felt right to him, doing otherwise would cause intolerable anxiety.

## Robert's story

Robert was a smart, normally well behaved and delightful ten year old boy. His parents brought him to the psychiatrist because they felt that something was not right with him as they noticed that he was showing some new behaviors, behaviors that sick people on TV displayed. Robert no longer wanted to share any cup or drink with another family member, something that was never a problem for him in the past. He did not want to eat food that he did not see being cooked to make sure it was not contaminated. This need of him set them up for some embarrassing situations when they went out to eat. The food items on his plate could not be touching each other. Robert was washing his hands approximately twenty five times per day. He was showering approximately four to five times per day and changing clothes, which put a burden on the family. Robert often became very anxious and angry when his family challenged him and put resistance to his demands. He became defiant or outright aggressive at times. Robert felt guilty and ashamed about his new behaviors and understood that he needed to change them but could not do it.

Robert had the compulsion of cleanliness with the obsession of contamination.

The examples presented above focused mainly on the effect that OCD had on the individuals with the condition, although one could have a glimpse of how OCD also affected family interactions amongst themselves and even with society at large. Family members incurred in significant expenses trying to help their loved one with OCD, often had to miss work due to dealing with the symptoms of the condition, or some times due to taking them to medical appointments. They did it all with abnegation and love, having as their only compass, the improvement of their loved one.

A perfect example of a public case of someone suffering with OCD is shown in the movie "The Aviator" starring Leonardo Dicaprio. This movie is an excellent depiction of the illness based on the life of Howard Hughes who was an intelligent and extremely successful man. The movie portrays how OCD slowly progressed over time until it overtook all aspects of Howard Hughes' life. Mr. Hughes' main compulsion was one of cleanliness, and his obsession was one of contamination. Due to his level of knowledge and intelligence his obsessions grew in complexity to the point that he considered that light, which is both wave and particle, was a pollutant that could contaminate him. He locked himself in a dark room where he stayed for months. The food had to be offered to him in a special way so that he could get it without having physical contact with anyone or with light. The movie seems to suggest that Mr. Hughes' condition had its origins in the relationship with his mother, who may have had some issues with contamination. This movie, although not scientific, is an accurate representation of the debilitating condition of OCD.

Another example of a movie showing people with OCD is the movie "Matchstick Men" starring Nicholas Cage. In this movie, the main character Roy has two main compulsions, cleanliness and symmetry. The accompanying obsessions are of contamination and order. The movie shows the severe anxiety that he suffers when he is exposed to the triggers of his compulsions and obsessions. The author also hints at how stressful events in Roy's life have an effect on his OCD illness. Like "The Aviator," this movie also seems to suggest that OCD is rooted in deep intrapsychic conflicts based on life experiences.

Finally, there is a movie named "Dirty Filthy Love". This movie shows the common comorbidity of OCD with other impulse control illnesses. The movie shows people with OCD and Tourette's Disease (a vocal and motor tic disorder often present with OCD), Thricotillomania (where the person pulls large areas of his/her own hair), and Depression. This movie is quite sophisticated regarding the theme of OCD as it deals with some of the concepts discussed in this book and some of the treatments. It is also an enjoyable movie about love, friendship and understanding.

There is a TV series titled "Monk" where the main character is a detective that has OCD. He presents multiple compulsions, the main ones being with cleanliness, symmetry and checking. He has the accompanying obsessions of contamination, order, and danger.

These movies and television show present an example of what it could be like living with OCD. A person who has this condition does not need to see these movies to know what OCD feels like and how it is to live with it, but for those readers who do not have the condition these movies could help them visualize it. In these movies the real time exposure of the spectator to the life of someone with OCD is for an hour or two. People with OCD usually live with this condition every day of their life, over many years. A lot of times some symptoms of the condition remain with the person through his/her life but in a manageable form. OCD usually has a fluctuating course, often with times of no symptoms or minimal ones.

# Epidemiology

There was a time when it was believed that there were very few people with OCD. According to multiple scientific studies, between two to four percent of the United States adult population is living with OCD. This makes approximately 4 million adult Americans. In addition, it is estimated that there are approximately 1.3 million children with OCD. The mean age of onset for OCD is in the twenties, although approximately one third to one half of the people with OCD have childhood onset. Twenty to thirty percent of these people also have Tourette's disease, a condition where people perform involuntary movements and noises that they wish they could stop, but feel driven to do. Some people are able to withhold these movements, usually for short periods of time, causing much anxiety. Then the movements and noises come back consecutively, usually worse than before, until the person goes back to baseline levels of movement and anxiety. It has similarities with OCD except that Tourette's disease only has a motor component or need to perform a movement, and it does not really have any obsessions or ideas associated with it. Fifty percent of childhood OCD cases may be familial, meaning that there are other family members affected by the condition. Seventeen to nineteen percent of parents of children with OCD also have the condition, and others have at least some of the symptoms. OCD is the ninth leading cause of disability in the United States of America (USA), and tenth in the world (in 1990).

The National Institute of Mental Health estimates that more than 2 percent of the USA population, or nearly one out of every 40 people, will suffer from OCD at some point in their lives. The disorder is more common than schizophrenia and bipolar disorder combined and it is equally common between men and women. It is a debilitating illness and many people develop co-morbidities with depression and substance dependence. People with OCD are at an increased risk of committing suicide.

OCD is categorized as one of the Anxiety Disorders. Anxiety Disorders are the most common mental illnesses in the United States of America (USA), affecting 19.1 million (13.3%) of the adult USA population between the ages of 18-

54 years of age. According to "The Economic Burden of Anxiety Disorders," a study commissioned by the Anxiety Disorder Association of America (ADAA) and based on data gathered by the association and published in the *Journal of Clinical Psychiatry*, anxiety disorders cost the U.S. more than 42 billion dollars a year. This is almost one third of the 148 billion dollars total mental health bill for the USA. More than 22.84 billion dollars of those costs are associated with the repeated use of healthcare services from those with anxiety disorders who seek relief for symptoms that mimic physical illnesses. People with an anxiety disorder are three-to-five times more likely to go to any doctor, and six times more likely to be hospitalized for psychiatric disorders than non-sufferers.

# Obsessions

The Diagnostic and Statistical Manual (DSM, created by mental health professionals in the USA to define and classify the mental or emotional afflictions), defines obsessions as recurrent, persistent thoughts, images, or impulses. People experience these obsessions as intrusive or inappropriate. These obsessions are not simply excess worries about real life problems like bills, marital or academic matters. They cause marked anxiety or distress. They must be recognized as products of one's own mind instead of implanted thoughts from an alien source. People who have other conditions and are delusional (they hold a false belief despite all evidence to the contrary) or psychotic (they lose contact with many aspects of reality and usually suffer from hallucinations) do not realize that their thoughts are products of their own mind, instead they can think that these thoughts are being introduced into them by some external entity by using telepathy, radio waves, or some other means.

## Compulsions

Compulsions are defined in the DSM as repetitive behaviors (e.g., checking locked doors, hand washing), could also be mental acts (e.g., counting, repeating words a certain number of times, praying), that a person feels driven to perform in response to an obsession or according to rigid rules.

## Traditional definition of OCD

The DSM defines OCD as obsessions or compulsions that are sources of marked distress. These obsessions or compulsions consume more than one hour a day, or significantly interfere with the person's normal routine or occupational and/or social functioning. These thoughts and acts are recognized as excessive or unreasonable (this recognition is not necessary in children). This is in contrast to the people that are delusional or psychotic. People that are delusional or psychotic do not realize that what they are doing does not really make sense.

# What is OCD?
# A biological theoretical approach

Of all the differences that we have with other animals in the animal kingdom, the most important one is the human development of a very large Cerebral Cortex, more specifically the Frontal Cortex. The Frontal Cortex occupies 3.5% of the total cortical volume in cats, 11.5% in monkeys and 30% in humans. It is involved in abstract thinking, future planning, and other high cognitive functions such as hypothesis generation and organization of information, which are mostly human capacities. This is what decidedly makes the major difference between humans and other animals. Other important functions of the Frontal Cortex are: inhibiting behaviors, shifting from one behavior to another, and motivation, which we share with other animals.

Nearly a century ago the father of psychoanalysis, Sigmund Freud, developed the Structural Model of the mind. In this model he described the Id, the Ego, and the Superego. The Id is the part of the mind that contains the basic drives, instincts or impulses. The Superego is the part that deals with our ideals and our morals, things that we have learned from our parents and from society. The Ego's task is to find a balance between reality, the primitive instincts or drives, and the Superego. The Ego is also our "conscious" self. This Structural Model of nearly a century ago is comparable to current knowledge of brain's function. The brain has a Limbic System, where our basic emotions and drives reside. This Limbic System is more primitive than the brain Cortex and is something that humans have in common with other mammals such as dogs and cats. This may be the homologous of Sigmund Freud's Id. Humans have a brain Cortex that deals with higher cognitive functions, such as planning, abstraction and moral matters. The brain Cortex may be homologous to the Superego. Sigmund Freud's observations of what he called the Ego are the result of the interactions between the brain Cortex and the Limbic System. The brain Cortex is evolutionarily more advanced and is constantly inhibiting and modulating the more primitive Limbic System. The outcome of the interaction between these two areas of the brain is what forms who we are. Sigmund Freud placed overwhelming importance on the sex drive. Libido is a very important drive but there are many others that are also very important. Drives or instincts are very primitive and have evolved to promote our survival. Before we can reproduce through sex we need to survive to the age of reproduction. Furthermore, there are many chances at sexual reproduction but life can be lost only once. Even with this singularity of life, there are individuals that will risk life for an opportunity at reproduction. Some mammals will place themselves at risk, but when the time comes, instincts having to do with survival at any given time are going to be stronger than other instincts. Among the many instincts are fear, hoarding, checking for symmetry and danger, cleanliness, sexual desire, and aggression (either predatory or defensive).

A lot of people think of OCD as a disorder of obsessions, and a secondary component of compulsions that are carried out in an attempt to alleviate these obsessions. The neurobiological theoretical approach argues the opposite. In this theory the compulsions develop first and then the person develops ideas or obsessions trying to explain these compulsions. These ideas can be as complex as the person's intelligence and knowledge.

OCD is the maladaptive urge to perform a genetically inherited sequence of behaviors that contribute to checking for danger, contamination, hoarding, and/or order. These behaviors are among the most commonly characterized as OCD and may be performed physically such as when locking a door or making sure the stove is off multiple times, or mentally such as counting and praying. They originate from areas of the brain that are more primitive than the Frontal Cortex, strongly stimulate the Frontal Cortex among other brain regions, and then the Frontal Cortex assigns thoughts or meaning to them. These genetically inherited sequences of behaviors are also called instincts.

Most people and animals get the urge to clean themselves after getting dirty, accompanied by a sensation of dirtiness. This urge or impulse originates in primitive areas of the brain where instincts originate. Then this impulse travels or stimulates the more evolutionarily advanced Frontal Cortex, which now will think about germs, radiation and other types of contaminants. The complexity of these thoughts is probably related to the level of education of the person. For example, a person that does not know about the existence of germs will not have an obsession about germs, the obsession will only be about dirtiness or cleanliness which is the purer, more primitive form.

In another type of compulsion and obsession the person has an urge to protect him/her self or others, accompanied by a sensation of danger or anxiety. The behavior and the accompanying sensation of anxiety are probably integrated as part of a response sequence in the brain. These behaviors, impulses or sensations most likely arise from primitive centers located in the inner layers of the brain and not by the higher centers of cognition located in the outer layers. These higher centers of cognition become involved afterwards giving meaning to or explaining these behaviors, impulses or sensations.

When the person has the urge to perform a behavior, if the behavior is not performed in accordance to the person's urges, the different areas of the brain will continue to be stimulated causing a sensation of increasing anxiety and urgency. This stimulation is usually episodic and slowly diminishes over time within each episode, although in the long term they can become more severe and of longer duration. Often these episodes are triggered by an event, exposure or as described by Konrad Lorenz (1982) and Niko Tinbergen (1951), a "releasing stimulus". A releasing stimulus is an event that unravels an automatic sequence of behavior. For example an itch unravels scratching. During this time the person performs the behavior keeping the anxiety from becoming overwhelmingly intolerable but at the same time making the behavior more automatic or procedural in nature. The more automatic or procedural the behaviors are, the easier they arise in the future.

## Procedural behaviors

Procedural Behaviors are behaviors, that once started, go through to completion, in a semi-automatic fashion, unless inhibited or shifted to another behavior by the Frontal Cortex. These may be learned or genetically inherited.

The Caudate Nucleus and Putamen are primitive areas of the brain. They are located in the inner layers of the brain and have many functions. Among these functions are the scanning of the environment and the performance of simple or automatic behaviors. There appears to be genetically imbedded information in these centers to recognize stimuli that represent danger such as fangs, claws, colorful insects, spiders, and movements. These stimuli are detected by the Caudate Nucleus and Putamen, which immediately recruit the attention of the Frontal Cortex to evaluate the stimuli and invest itself in solving the situation, or deem the stimuli as non-threatening. These nuclei also have the capacity to immediately start a response to put the individual out of the possible danger, even before the Frontal Cortex has time to evaluate the situation. This explains how a person can surprise or scare us and cause what some consider the funny startle response. The Caudate Nucleus and Putamen started a survival response before the Frontal Cortex determined that it was a person's attempt to be funny and not an actual threat.

Other functions of the Caudate Nucleus and Putamen are the inhibition of impulses or information from other areas of the brain and from the environment, to the Frontal Cortex. Inhibition of unimportant information from reaching the Frontal Cortex allows the Frontal Cortex the freedom to perform the higher functions that it is capable of doing. The Caudate Nucleus and Putamen are also involved in assisting in the coordination of most of the behaviors that we perform. We perform many behaviors in a semiautomatic way. These behaviors are primarily managed by the Caudate Nucleus and Putamen while the Frontal Cortex is involved in thinking and decision making. If the behavior is simple and well known, once a decision is made by the Frontal Cortex about what will be done, the behavior is managed by other areas of the brain including the Caudate Nucleus and Putamen. One example of this type of simple behavior is walking. The Frontal Cortex makes the decision of the direction one should take while the actual walking is handled by the Motor Cortex and assisted by the Basal Ganglia (composed of Caudate Nucleus, Putamen, Globus Pallidus, Substantia Nigra) and the Cerebellum. This leaves the Frontal Cortex free to perform other less automatic functions and tend to new issues that may arise within the individual or surrounding environment.

Learned Procedural Behaviors are behaviors that while being learned, require full Frontal Cortex engagement due to their complex nature and degree of self-monitoring required. Once these behaviors are practiced over and over, they become procedural in nature and are performed mostly by more primitive areas of the brain such as the Caudate Nucleus and Putamen. At this point the Frontal Cortex is minimally engaged. The more a behavior is practiced, whether physically or mentally, the more procedural and automatic in nature it becomes. For example, Olympic divers often visualize the entire diving process from the first lift from the board to the clean entry to the water. This has proven to help them perfect the real behavior of diving into the pool.

One example of a learned procedural behavior is driving a car. When driving a car for the first time, especially if the car has a manual transmission, the driver is fully engaged and focused in the process, thinking little about other things. Having driven enough times it becomes procedural behavior or "second nature," requiring much less focused attention. As a matter of fact, drivers engage in abstract thinking and future planning (functions performed by the Frontal Cortex) when driving, because the Frontal Cortex has become almost completely disengaged from the procedure of driving. Although not recommended, it is common to see drivers in a passing car who are singing, grooming themselves, reading, and even texting while driving. Maybe the following has happened to you? One day you changed your routine about where you were going. Instead of going to work like you did every morning you had an appointment in another location. If you disengaged your Frontal Cortex too much from your driving you may have ended up at work, instead of at your new appointment. It is almost as if your car took you on tracks to where you usually drove. Driving to work had become a procedural behavior and was happening almost automatically, unless more conscious thought was put into changing it. There have been people that usually do not take their children to daycare, but one day they had to. Because of the behavior of going to work has become procedural in nature, they drove to work overlooking that they still had the sleeping child in the car, that they were supposed to drop off at daycare. They completed their routine of going to work, leaving the child in the car with dreadful consequences.

Genetically Inherited Procedural Behaviors are different from the Learned Procedural Behaviors. Genetically Inherited Procedural Behaviors are behaviors that do not have to be learned. They could also be called instincts, drives or impulses. If one makes an analogy between the mind and a computer, one could say that animals, humans included, are born with a Windows Operating System and through life one installs a lot of software, which is the learned information. The Genetically Inherited Procedural Behaviors would then be analogous to the Operating System in a computer. As stated before they could also be called instincts. These behaviors can be accompanied by strong and primitive emotions such as a sense of urgency, danger, anxiety, pleasure, ecstasy…, and can very strongly stimulate or send signals to the Frontal Cortex. Their emotional intensity goes hand in hand with their evolutionary and adaptive value or importance. With this strong engagement or signaling to the Frontal Cortex, the Frontal Cortex becomes fully engaged in finding a solution to a situation. The Frontal Cortex is then able to inhibit the behavioral response if it is not appropriate under the circumstances, and/or alter it to a more appropriate one. There are times when the primitive areas where these behaviors are embedded become inappropriately activated, sending a signal to perform a Genetically Inherited Procedural Behavior that is not appropriate under the circumstances. Two examples of inappropriately activated behaviors and classic examples of OCD are: a person who feels compelled to wash his/her hands multiple times despite "knowing" the hands are clean; and a person who can not resist collecting and saving items which are not needed and for which there is no space. This signal strongly engages or stimulates the Frontal Cortex to deal with a need that is not based on reality but on what that the person feels is urgent. The Frontal Cortex assigns meaning to this stimulation, such as having germs in the case of the compulsive cleaner; or "saving for a rainy day" in the case of the compulsive hoarder. Still the Frontal Cortex is able to recognize the inappropriateness of the stimulation and of the idea, but it has difficulty stopping it, as well as the behavioral response and anxiety associated with it. This stimulation and behavioral response is OCD. In other words, OCD is an illness that happens when Genetically Inherited Behaviors, that are adaptive in nature, become deregulated, or inappropriately activated.

# Ethology

Ethology is the study of animal behavior. Despite the fact that the human brain is different from other animals, there are many similarities, and this is why Ethological studies are critically important to understand human behaviors. A lot of human behaviors happen in animals in a more rudimentary form and without the confounding factors of a highly complex Frontal Cortex and Social System.

A central concept in Ethology is the "Fixed Action Pattern" defined by Konrad Lorenz as an innate and adaptive behavior sequence that is actualized by a Releasing Stimulus. Also discussed by Dr. Dan J. Stein and collaborators (Comprehensive psychiatry 1992). A Releasing Stimulus, as stated before, is a trigger that prompts a human or animal to perform an automatic or procedural behavior. For example, physicians used to slap newborns bottoms to get them to take their first breath. Upon being slapped newborns immediately performed the act of crying which involves coordinated movements of the diaphragm, abdominal, inter-costal and neck muscles, and closure of the larynx in an automatic or procedural fashion. The Releasing Stimulus for this sequence of behavior and movements was the slap, and the procedural behavior was an audible outburst, or cry. Some of these behavior sequences are expressed at birth like the example just given, the pecking of a newly hatched chick trying to break the egg, and the sucking reflex (when a finger, a bottle, or a nipple is placed in a baby's mouth, the baby immediately starts to suckle on it). Others are expressed later in life, like the urge to fly in birds (although they need practice to accomplish the actual flying), sexual urges, and courtship behaviors. There is a fish that has been extensively studied called the Three-spine Stickleback. A section of this fish turns red once it reaches sexual maturity. Male fish of this species will attack other males for the privilege of mating. In fact, it has been discovered that they will attack anything that is red. The color red is the releasing stimulus for the aggressive behavior in this animal. This Fixed Action Pattern is seen in all members of the specie, even when they are raised in isolation. Once initiated the Fixed Action Pattern usually continues to completion unless stopped by the Frontal Cortex in more evolutionarily advanced animals. The responsiveness of an animal to releasing stimuli is determined by a variety of internal and external factors. Learning may also play a role in determining whether stimuli actually become releasers and in orienting the execution of a fixed action pattern.

There is a fable about a scorpion that needed to cross a creek. Since the scorpion could not swim he walked along the shore until he found a frog. He asked the frog if the frog could carry him to the other side of the river. The frog refused secondary to his awareness of how dangerous scorpions can be. The scorpion begged and convinced the frog to take him over by promising that he would not sting him. The frog finally agreed. When they were in the middle of the creek something happened and the scorpion stung the frog. As they were both drowning the frog asked the scorpion why he had stung him. The scorpion sincerely apologized stating that he could not stop himself, that it is his nature. This is an example of a fixed action pattern being carried out by the scorpion, despite its lethal consequences for both.

# Most common compulsions and obsessions

The following are amongst the most common compulsions and obsessions observed in OCD.

Cleaning/grooming rituals are usually paired with ideas of contamination or cleanliness. The idea of cleanliness is important in this theory since this disorder of excessive cleanliness has been described even before bacteria and germs were known. The term "scrupulosity" dates back to the twelfth century. At that time the diagnosis was not called OCD but a disorder of Excessive Scrupulosity, which was also associated with excessive religiosity. People, in addition to keeping themselves immaculately clean, also feared that their behavior was so defective and unfit to be presented unto God, that he was not going to accept it. During this time there were no doctors and barbers performed surgeries with little attention to cleanliness, especially hand washing, leading to infections and an extremely high mortality rate from the surgical procedures. People with Excessive Scrupulosity maintained themselves excessively clean despite the world being ignorant about the existence of germs. So they did not have fear of contamination, they had fear of dirtiness. Then, like today, some people had to wash each area of their body a hundred times, taking hours in the shower. Others washed their hands up to a hundred times per day or more. These extreme behaviors over time can lead to dermatological conditions such as skin exfoliation and ulcers. A lot of people with OCD have to use creams and lotions to treat their skin.

Placing objects in order, and/or repeating things or ideas, are accompanied by their usual obsession of perfectionism or symmetry. Some people arrange their clothes by colors or by fabric type. Others have to tap on each side of the door ten times every time that they go through. There are others that have to count in odd or even numbers until the reach one hundred every time they feel the need to.

Hoarding is usually accompanied by the idea of saving. It is difficult to get rid of unnecessary things with the idea that they might have some use in the future. Some people collect items that at times become very valuable. If the person is organized and collecting does not affect him/her in a negative way, it could be an adaptive or positive thing. Others buy a hundred brooms despite having five hundred at home because they were on special and they could not pass without buying them. Some people pick up garbage from the streets. Sometimes they drive by some piece of garbage, resist the urge to stop and pick it up, and then come back from home to pick it up. They could not tolerate the anxiety created by suppressing the urge to pick it up. Some people gather so many items for which they really have no use, that they literally put themselves out of their homes.

The behavior of checking if the doors are locked or if the stove is off is commonly associated with preoccupation about danger or violence. This preoccupation is about undesired violence, not predatory violence. Predatory violence is the one shown by a cat when it is hunting a mouse. The cat at this time is calmed, "calculative," and feels well. The undesired or emotional violence would come from the mouse trying to defend itself. It will do it in a panicky and disorganized way. Predatory violence is the one done by psychopaths when they victimize someone. These two types of violence are very different in almost all aspects and its discussion is not part of the scope of this book. It is probable that predatory violence also has a compulsive and primitive nature, although it is usually not included as one of the classic OCD symptoms. Another example of preoccupation with undesired violence is the one given in the beginning of the book, where the woman had to check that the doors were locked for up to four hours after having opened them.

In the animal kingdom there are many examples of behaviors that resemble the mentioned compulsions and obsessions; orderliness, hoarding, checking for danger, and cleaning/grooming.

African Weaver's nest

Birds build nests that are specie specific. They build the same type of nest as all other birds of their species, even if they have been raised in isolation and have never seen a nest before. This type of ability requires a genetically inherited pattern of behaviors that the bird does not have to learn. This is an adaptive strategy that requires symmetry and a particular procedure. In the African Weaver's nest in the picture, the main evolutionary difference has been the development of the use of fresh leaf fibers to build the nest, as opposed to dry twigs or other materials.

This nest building behavior could be inappropriate or maladaptive if a bird was so particular about the way it built its nest, that the nest was not ready for when the bird needed to lay the eggs. Such a bird would have the compulsion to build the nest a certain way that could interfere with its ability to tolerate minor and/or irrelevant deviations. It would take too long for this bird to build "the perfect nest", thus preventing it from finishing the nest on time to lay the eggs.

In humans the capacity to appreciate symmetry and differences in the environment is of extreme importance. It helps with orientation and also with detecting danger because of changes in the environment. This is especially important since humans lack the sensitivity in some of the senses that other animals have, like smell and sight. For pre-historic humans it would have been a lifesaving ability to notice that things were not like they left them when they came back to the cave. This could have meant that there were animals or other humans in the cave which could have been dangerous to them. This sense of symmetry also helped them to track down animals during hunting and allowed them to build effectively. It is well known that having a disorganized environment causes anxiety and mental disorganization in most people. An advice that is probably positive for most people is to maintain their environment organized, since it causes a state of well being. A tendency towards symmetry seems to be a genetically ingrained behavior. Furthermore, there are studies that indicate that people tend to judge beauty by degree of symmetry. Humans unconsciously notice that the height of both eyes is the same, as well as the proportion between height and other bodily features.

In some people this need for symmetry gets out of control. They need to spend hours aligning a few books by size, to the point of getting specialized tools to get exact measurements. By spending so much time doing this behavior, which they know does not really make sense, they neglect their other responsibilities. It often costs them their jobs and/or relationships.

North American Gray Squirrel

A prime example of animal hoarding is the squirrel. When a squirrel finds food there are two things that it will do with it; either eat it or hoard it. Usually if they do not eat it right away, they accommodate it in their large cheeks and bury or hoard it somewhere. This is very adaptive behavior because when the harvest season is over and there is no more food, the squirrel can go around and find some of the food that it has buried or "saved".

Now, with hamsters it is slightly different. The hamsters that are sold for pets have never been in the wild. They probably have never gone hungry. But they also have large cheeks where they hoard food, and they use them. They fill their cheeks with food and then bury it around the cage. People that do not know about this hoarding behavior of hamsters and feed them every time their plate is empty, end up with a cluttered cage full of buried or stored food.

The behavior of the hamster is the same as that of the wild squirrel. Because the hamster's living situation is not the same, the continuation of the behavior (which is a Genetically Inherited Procedural Behavior) is maladaptive. The hamster just does it because it is deeply ingrained in its brain, "it is their nature".

People with OCD hoarding type have the problem that the hamsters have. They do a normally adaptive behavior but they do it excessively, during inappropriate circumstances, making it maladaptive. They save things of which they have too many of, do not have space for, or just do not need. This behavior creates severe problems for them, they wish that they could stop doing it, and are unable to. For a lot of people though this is one form of OCD that is commonly "egosyntonic". Egosyntonic means that they feel fine about it or do not think that they have a problem. Because of this they do not seek help and often live alone because other people can not tolerate their lifestyle. In many cases they seek help because they are forced or coerced by other people, usually their loved ones or authorities.

American Goldfinch feeding

In prey animals, checking for danger is a very prominent behavior. Checking for danger is adaptive because the individuals that do not do it may end up being someone else's meal. Birds in bird feeders are often unable to eat because of the compulsion to check for danger every time that they are about to lower their heads to peck at the food. Sometimes they fly away and return several times before eating anything. If this behavior became excessive and interfered with their need to feed, it would be maladaptive.

In humans as in all animals, checking for danger is a necessity for survival and it is highly adaptive. When checking for danger becomes excessive such as Pam's case from the beginning of the book, checking for danger itself can become dangerous. If Pam had not had her husband who supported and provided for her, she would have been in great danger. Her ability to provide for herself was greatly impaired secondary to her need to check that the doors were locked repetitively. Remember that she spent four years without leaving her home.

White German Shepherd licking repeatedly

Many mammals lick their fur regularly to groom it, when they are dirty, and when they are injured. This is adaptive behavior since it has multiple functions. When an animal licks a wound it clears the area by removing foreign bodies, dead tissue, and hair. The saliva of many animals also contains lytic enzymes that destroy harmful bacteria. The act of licking also deposits the animal's normal flora (normal bacteria) in the wound. These are usually less harmful to the animal and also help eliminate harmful bacteria because they compete for the available resources, such as food and space. This is a very adaptive behavior, but the animal is not aware of any of this. It does not even know about bacteria but they still do the behavior consistently. They do it because it is a Genetically Inherited Procedural Behavior.

What happens when an animal does this behavior too much? Excessive licking, to the point where a dog causes damage to its own skin, is a condition that in many cases has psychological or emotional causes. It is a so-called psychogenic disease. Breeds most often affected include: Doberman Pincher, German Shepherd, Great Dane, Irish Setter, and Labrador Retriever. Sometimes dogs and other animals start licking their fur and do not seem to be able to stop. They often do it without having any pre-existing injury. They persevere at this behavior and do it for extended periods and multiple times per day. The condition is called Acral Lick Dermatitis. This can lead to severe ulcerations and cellulitis (inflammation of the skin). The condition is worsened by stress but it also appears on dogs that live under the best of situations. It is more common in certain breeds, pointing to a genetic predisposition. Acral Lick Dermatitis is very difficult to treat with the conventional veterinary treatments. It has been treated in multiple ways including steroid injections, antibiotics, antihistamines, and changing the dog's environment trying to decrease their stress, with poor results. Now several studies have shown robust responses to Clomipramine (Anafranil), Fluoxetine (Prozac) (Goldberger and Rapoport 1990), Citalopram (Celexa) (Dan J Stein, et al 1998). These are medications known to be effective in the treatment of OCD, other anxiety disorders, and depression in humans.

This is very similar to the people that have OCD with the compulsion of cleanliness. Nowadays this compulsion is accompanied by the obsession of contamination. In the past, before humanity knew about the existence of germs, humans were more like the above dogs, they just cleaned themselves because they were dirty. Sometimes they also paired it with religiosity, which is what was popular at the time. Humans can also cause severe skin damage due to excessive cleaning of their skin. Sometimes the act of cleaning takes so long that they have to quit school or work.

I once met Martha, a thirty four year old woman, who had been having problems with excessive cleanliness. Her condition waxed and waned over time, but had lasted close to four years by the time I met her. By the time I met her, she was showering approximately four to five times per day and each shower took her approximately one to two hours. She needed to scrub each little part of her body in a circular motion for approximately five minutes. Failure to do so caused severe anxiety and the feeling that she was dirty which made her do it even longer. Basically her entire day was spent taking showers. Although she was a very bright woman she could not work or be independent and depended on her family to support her. She had moderate hand skin exfoliation because of the excessive washing. Her father and sisters were physicians and dentists. Her family's success worsened her feelings of disability or incapacity. In addition, Martha was suffering from severe depression. She was on multiple medications and had seen many different physicians for her conditions, with only partial improvement at best. At the time that we met she was being treated by an OCD expert somewhere else, which made our interaction very limited. Later on I learned she had committed suicide.

# Neuroanatomy and Physiology

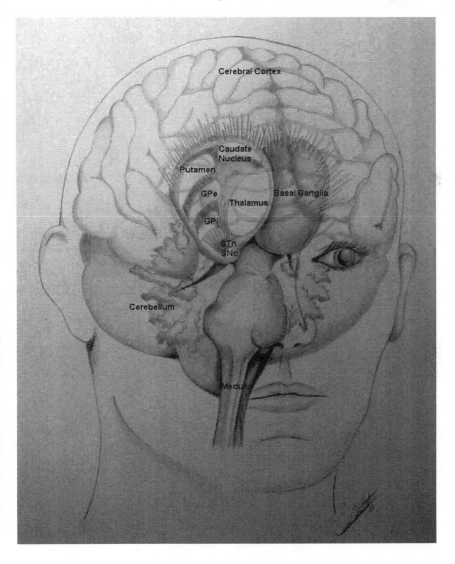

Brain neuroanatomy

As part of discussing the neuroanatomy and physiology of the brain, we need to review some basic structures and definitions. The Basal Ganglia depicted in the figure is composed of the Caudate Nucleus, Putamen, Globus Pallidus (GP), and Amygdala. The Thalamus is not part of the basal ganglia but it is very close to it as seen in the picture. The Corona Radiata becomes the Internal Capsule and has tracts that go from the brain to the body and from the body to the brain. The term Striatum refers to the Caudate Nucleus and the Putamen together. Information coming from the body goes through the Thalamus and the Basal Ganglia, to other areas of the brain. Information going from the Cortex to the body, and the Cortex to other areas of the brain, also passes through here.

There are also Cortical Association Pathways that go from cortical area to cortical area without passing through the Basal Ganglia-Thalamus Complex but are not going to be discussed in this book. Notice the Cerebellum and the prominent tracts it sends into the Basal Ganglia-Thalamus area. The importance of this will be discussed further along.

The Thalamus, as stated before, serves as a relay station where information is sent to different areas of the brain. The Thalamus also has other functions that are beyond the scope of this book.

The Basal Ganglia (the Striatum being part of it), as noted earlier receives information from the body going to the cortex, from the cortex going to the body, and from the cortex going to other areas of the cortex and the brain. Among other things The Basal Ganglia serves as a gating mechanism working in combination with the Thalamus. The Basal Ganglia filters out information getting to the Prefrontal Cortex, and also enhances other information. The Basal Ganglia, as a more primitive structure than the Frontal Cortex, receives information from the environment all the time. When The Basal Ganglia gets information that reaches a certain threshold of relevance, it alerts the Frontal Cortex, so that a higher level of evaluation can be performed. If upon evaluation by the Frontal Cortex, the information is deemed unimportant, it could again be relegated to being monitored by the Basal Ganglia until a change is detected, that might again warrant evaluation by the Frontal Cortex. Examples of information that have a high likelihood of being sent to the Frontal Cortex for evaluation are unexpected movement, changes in intensity of noise, novel things in the environment, things out of place, or anything that could be a threat to the individual. Things like the noise from the air conditioner, vibrations, and other noises probably will not come to the attention of the Frontal Cortex unless we consciously decide to pay attention to them. These noises and vibrations are continuously monitored by the primitive structures, Thalamus and Basal Ganglia. If there was a change in their intensity or frequency, or if the air conditioner suddenly stopped working, the Frontal Cortex would be recruited or informed, and we would become conscious of the activity of these vibrations and noises.

The Basal Ganglia is also involved in procedural learning, which was defined earlier. Different regions of the Basal Ganglia receive input from Paralimbic Isocortical Areas, Associative Neocortical Areas, and Limbic Areas. These areas of the brain cortex are evolutionarily more advanced than the Basal Ganglia and the Thalamus. They are more involved in non-automatic activities, or thoughtful activity (thinking). Circuits involved in motor programming (programming of movement) travel through the Putamen, which is part of the Basal Ganglia. The Basal Ganglia receives input from cortical areas involved in thinking and thoughtful movement. Over time and with practice, the Basal Ganglia can take over those actions that initially needed thinking or involvement of the cortical areas, and do them with only minimal involvement of the cortical areas. At this point they become Learned Procedural Behaviors.

Excitatory or stimulatory projections use mainly glutamate as a neurotransmitter. Inhibitory or brain slowing projections use mainly GABA as a neurotransmitter. Other naturally occurring neurotransmitters in the brain are Dopamine, Serotonin, Substance P, Enkephalins, Acetylcholine, Opioids, Cannabinoids, and many others that continue to be discovered. These neurotransmitters are important because they are probably all affected to some degree in people who suffer OCD since these chemicals interact in synchrony or balance with each other. When the amount of one of these naturally occurring chemicals fluctuates, either because of illness, injury, and medications, the others adjust accordingly, balancing brain functions. Currently, scientists have developed medications that exert their effects through mostly Serotonin and Dopamine, and more medications will be developed in the future to assist with OCD treatment.

The Orbito-Frontal Cortex (OFC) (Frontal Cortex located near and above the eyes) maintains extensive connections with the Amygdala, Hypothalamus, Temporal Lobe, Basal Forebrain, autonomic centers in the Brain Stem, and motor centers. The Amygdala is involved in the development of strong emotions such as rage or horror. It also gives emotional significance to memories and experiences. The Hypothalamus is involved in activating the Autonomic Nervous System; increasing heart rate, dilating the pupils and bronchioles in the lungs, and liberating adrenaline for the fight or flight response. The Temporal Lobe, including the hippocampus, is where memories or information are stored. The autonomic centers in the Brain Stem raise the blood pressure, decrease or increase bowel motility, and increase heart rate, among other functions. The motor centers produce skeletal muscle movement. These are muscles that are connected to the bones and are moved voluntarily, as opposed to the muscles in the gut that move on their own or involuntarily. The Orbito-Frontal Cortex (OFC), Thalamus, and Basal Ganglia share connections with all of these structures. Since it is believed that OCD is explained by overactivity and deregulation of many of the structures just mentioned, it is understandable why a person with OCD has strong emotional responses as well as autonomic nervous system responses (increased blood pressure, vomiting, diarrhea, sweating, tachycardia, and shortness of breath). People with OCD as well as those with other Anxiety Disorders, commonly have "physical symptoms" associated to the anxiety.

# Other medical disorders associated with OCD symptoms

There have been multiple medical conditions that have been associated with an increase incidence of OCD symptoms. Scientists started noticing that some people who previously had no OCD symptoms, developed certain medical conditions and started having OCD symptoms, often quite abruptly, after developing the medical condition. Scientists were able to link the OCD symptoms with the development and progression of other medical conditions such as Sydenham's Chorea and Postencephalitis Parkinsonian Syndrome. In other words, there are some general medical conditions that may include symptoms of OCD. This does not mean in any way that people with an OCD diagnosis will develop Sydenham's Chorea or Postencephalitis Parkinsonian Syndrome, but that people who develop these conditions might develop OCD symptoms.

OCD is an illness of minor chemical and structural changes in the brain. A brain structural change refers to a relatively large change in size or shape of an area of the brain. These changes, if large enough, can be picked up or measured relatively easy in neuroimaging studies such as X-rays, MRI, and CT-scans. The chemical changes in the brain with OCD are small, but they are enough to change how the brain works, even though the size and shape of the different brain regions remains almost the same. With the technology available in most hospitals today, small changes in size or shape can not be detected, or if detected are not meaningful. In order for physicians to know what these changes mean, they would have to compare their findings against brains in the general population. Neuroimaging studies for OCD are attempted and done only under research conditions. Doing neuroimaging studies during evaluation and treatment by a physician in the community is still not useful. The diagnosis of OCD is made through an interview, not through laboratory tests. Other medical disorders associated with OCD symptoms, like Post Encephalitis Parkinsonian syndrome, are caused by large structural changes that cause OCD symptoms when the changes are in areas of the brain that have been linked to OCD.

## Postencephalitis Parkinsonian Syndrome

The Basal Ganglia is a brain structure that among other functions, deals with the coordination, modulation or inhibition of brain impulses coming from the brain to the body, from the body to the brain, and from within some regions of the brain to others. In Postencephalitis Parkinsonian Syndrome the Basal Ganglia is affected. This is a type of Parkinson that developes after encephalitis (or brain inflammation). There was an epidemic of encephalitis from 1918 to 1926, when patients, after developing encephalitis, were left with a permanent Parkinsonian syndrome (see the movie "Awakenings"). They had symptoms typical of Parkinson's disease; an involuntary movement disorder in which the person slowly develops tremors, rigidity, slowness of movements, flattening of affect, Dementia, and progresses to death over many years. Many of these people also developed symptoms of OCD. This was one of the first times that OCD was thought of as a brain disorder with a biological basis as opposed to poor parenting, upbringing or a character or personality defect.

## Huntington's Chorea

Huntington's Chorea is a genetically transmitted condition carried by chromosome 4. It is a condition where the Caudate Nucleus degenerates and dies. This Nucleus is part of the Basal Ganglia, therefore coordination, impulses, or signals from the brain to the body, body to the brain, and brain to brain are affected. In the initial stages it is comparable to having an electrical short circuit. These patients develop a severe involuntary movement disorder with undulating body and alternating limb movements. They develop severe personality changes with aggression, depression, mood lability, and many also develop OCD symptoms. The disease ultimately leads to dementia and death.

## Bilateral Necrosis of the Globus Pallidus

Bilateral Necrosis of the Globus Pallidus is a condition where the Globus Pallidus dies. The Globus Pallidus is part of the Basal Ganglia and it is believed to be the area of the Basal Ganglia that exerts the final inhibitory or modulating action of the Basal Ganglia over the Thalamus. When there is necrosis (death of tissue), OCD symptoms develop.

### Tourette's Disorder

Tourette's Disorder, has a high comorbidity with OCD, and vice versa. In other words, a lot of people with Gilles de la Tourette's also develop OCD. Gilles de la Tourette's is a condition where the person develops motor and vocal tics. Motor tics could be grimacing, blinking, and arm or leg movements. Vocal tics could be throat or nose clearing noises, hissing, and even words, including the famous although uncommon foul words that people say involuntarily (see the movie Maze). Gilles de la Tourette's Syndrome is believed to be an illness where the Basal Ganglia is affected, as in OCD.

### Pediatric Autoimmune Neuropsychiatric Disorders Associated with Streptococcus (PANDAS)

Pediatric Autoimmune Neuropsychiatric Disorders Associated with Streptococcus (PANDAS) is a controversial diagnosis at this time. Children some times develop OCD following a sore throat infection caused by Group A Beta hemolytic Streptococcus. This infection might not even be remembered by the child or the parents at times, but when blood tests are done, there are high levels of antibodies against these bacteria. This indicates that there was a recent infection and the body produced the antibodies to fight it. Some investigators have treated the OCD condition with antibiotics against the bacteria and/or by doing plasmapheresis. Plasmapheresis is a process by which some substances such as antibodies produced by our body against the bacteria can be removed from the body with some success. It is theorized that when the body fights against this bacterial infection, the body produces antibodies to destroy the bacteria. These antibodies then continue circulating in the blood stream and can attack some of the individual's own organs, which have "markers" (particles, or chemicals) that resemble those of the bacteria causing a cross-reaction. In the case of some people who developed OCD after the infection, the Basal Ganglia was found to be inflamed or damaged, probably by antibodies produced to help kill the bacteria. This is the same theory that explains rheumatic fever, rheumatic arthritis, and rheumatic heart disease. In these conditions the body also develops antibodies to fight the infection and then those antibodies attack the heart, joints, and other areas. Following a bacterial infection in which the Basal Ganglia is affected, the person may develop OCD and/or an accompanying involuntary movement disorder such as Sydenham's Chorea characterized by uncontrollable undulating movements. Sydenham's Chorea is a prime example of PANDAS.

## Hemiballism

Hemiballism is another movement disorder where it is believed that there is damage to the Caudate Nucleus or other parts of the Basal Ganglia. The person develops flailing movements of the limbs. It is also associated with OCD.

## Orbito-Frontal Meningioma

Orbito-Frontal Meningioma is a malignant cancer located in the Orbito-Frontal Cortex. People develop dementia and eventually die. Initially, in some cases Orbito-

Frontal Meningioma is associated with OCD as well. The Orbito-Frontal Cortex is another one of the areas that has been found to be altered in people with OCD and forms part of the malfunctioning circuitry in OCD.

## Psychosurgery

Since OCD is a brain/biological disorder, physicians have successfully produced strategically located lesions through techniques called psychosurgery to treat severe cases of OCD that have not responded to less invasive forms of treatments. Psychosurgery is an uncommon treatment which is performed in a few very specialized centers around the world. It is mostly performed after the person has been treated with psychotherapy, medications, Electro Convulsive Therapy, and various other treatments with no positive result. The four types of psychosurgeries most often performed are: Subcaudate Tractotomy, Anterior Cingulotomy, Anterior Capsulotomy, and Limbic Leucotomy. However, new procedures are being developed at the different treating and research centers.

To help decrease the symptoms of OCD, neurosurgeons interrupt white matter tracts between the Orbito-Frontal Cortex and Sub-Cortical structures, in a procedure called Subcaudate Tractotomy. White matter tracts are collections of axons of neurons. Neurons are one type of several different brain cells. The axons are the tails of the neurons and they serve the purpose of transmitting impulses or information between the neurons. The neurosurgeon interrupts these tracts and the two brain areas connected by them can not communicate or do so in a lesser way, depending on the extent of the interruption.

When psychosurgery is the course of action, Anterior Cingulotomy is the most common procedure used to treat OCD in the United States. A lesion is produced by neurosurgeons in the anterior aspect of the Cingulate Gyrus for the purpose of disrupting connections between the Anterior Cingulate Gyrus and the Caudate Nucleus. The Anterior Cingulate Gyrus is another brain structure believed to be implicated in the development of OCD. After the lesion, a reduction of Caudate Nucleus size has been observed linked to a decrease in OCD symptoms.

In the Anterior Capsulotomy the neurosurgeon disrupts tracts between the Frontal Cortex and the Thalamus, which leads to decreased stimulation of the Frontal Cortex thus decreasing OCD symptoms.

Limbic Leucotomy is a surgical procedure that combines a Subcaudate Tractotomy and Anterior Cingulotomy. In this procedure, the surgeon decreases OCD symptoms by interrupting white matter tracts communicating the Sub-Cortical structures and the Cortex, and at the same time the anterior aspect of the Cingulate Gyrus and the Caudate Nucleus, ultimately decreasing the size of the Caudate Nucleus.

In addition to the more invasive treatments mentioned above, scientists and physicians are currently exploring the use of electrodes in the brain. Information in the brain is carried within the cell in the form of an electrical impulse. Once that impulse reaches the end or tip of one cell, it secretes a chemical that reaches the next cell which in turn is activated. That cell then transmits information electrically within itself until the information reaches its tip. It then secretes a chemical to pass the information on to another cell and the process repeats itself. Researchers can strategically place electrodes in the brain, then stimulate those electrodes causing a disruption of the cells transmissions, or an interruption in cells function without killing the cells. This research is promising because electrodes could be removed or "turned off" allowing the original cells to again perform their function.

# Neuroimaging and other studies

Nowadays we have technology that the pioneers in the Mental Health field did not have available when they started developing their theories. Neuroimaging has been a breakthrough in the study of the brain as it has allowed researchers to study the human brain without having physical access to it (without opening the skull). Neuroimaging is sometimes divided into two categories; Structural and Functional neuroimaging. Structural neuroimaging gives a picture showing the anatomy of the brain and is used for detecting structural abnormalities such as tumors, bleeding, and strokes. Traditional X-Ray, Magnetic Resonance Imaging (MRI), and Computer Tomography Scan (CT-Scan) are several examples of structural neuroimaging. Szeszko et al 2004, Kim et al 2001, Gilbert et al 2001, Giedd et al 2000, Szeszko et al 1999,, Rosenberg et al 1998, Jenike et al 1996, Bartha et al 1998, Robinson et al 1995, Scarone et al 1992.

Functional neuroimaging studies allow researchers to see the working brain. It is possible to measure the relative amount of oxygen that a brain is using in a given area at a given time, the amount of glucose (main source of energy of the brain) being used, and other chemicals. These studies are extremely important because they allow monitoring of the live brain at work. They can measure differences in brain function between different people, or in the same person while performing different tasks, and/or over time. In the future it will be possible to detect changes in the brain which could be affecting brain function before they are large enough to cause structural changes. Once structural changes are present they are most likely harder to reverse. Functional Magnetic Resonance Imaging (fMRI), Positron Emission Tomography (PET), Single Photon Emission Computed Tomography (SPECT), and Magnetic Resonance Spectroscopy (MRS) are a few examples of functional neuroimaging equipment and techniques (Saxena, Brody, Maidment, et al 2004, Saxena et al 1999, Schwartz et al 1996, Perany et al 1995, Rauch et al 1994, Nordahl et al 1989, Swedo et al 1992 and 1989, Baxter et al 1992, 1988 and 1987, Busatto et al 2001, Lucey et al 1995, Harris et al 1994, Rubin et al 1992, Hoehn et al 1991, Machlin et al 1991, Adams et al 1989, Rosenberg et al 2001, Fitzgerald et al

2000, Bartha et al 1998, Baxter et al 1996, Ebert et al 1996, Pujol et al 1999, Rauch et al 1997, Breiter et al 1996).

It is necessary to acknowledge that in the field of Psychiatry or Mental Health in general, neuroimaging technologies are mainly used for research purposes at this time. They are not useful for the clinician in the evaluation or treatment of patients. However, neuroimaging studies are clinically important as they are used to evaluate whether the person has other conditions such as tumors, strokes, arteriovenous malformations, and Multiple Sclerosis to name a few. These conditions if suspected or diagnosed are further evaluated and treated by other medical specialists.

Other technologies, such as the Electroencephalogram (EEG), study the electrical activity of the brain and it is most often used clinically in the evaluation of seizures.

Transcranial Magnetic Stimulation (TMS) changes brain activity through magnetic pulses. It has been used to study the brain behavior in healthy people and in people with OCD after the magnetic stimulation. It has also proven effective in assisting in the treatment of other psychiatric conditions such as depression.

Of all of these technologies the most useful ones for research purposes are the functional metabolic technologies including PET, SPECT and fMRI. The structural or anatomic changes in OCD are usually very small and are hard to detect with current technology. Chemical changes are proving to have a greater significance, and functional technologies can detect these changes, although as mentioned before, only under research conditions. The functional technologies measure how the brain is actually working. Researchers already know with quite detail the anatomy of the brain and now they need to know more about how it functions. These technologies are invaluable for this purpose.

Baxter et al. and Benkelfat et al. designed studies where they examined both, people with OCD and without the illness. The researchers performed functional studies in both groups of patients during resting state. They also measured patient's brain activity after symptoms were provoked. Researchers found, in people with OCD, increased metabolism or consumption of energy in the Orbito-Frontal Cortex (OFC), Anterior Cingulate Gyrus, Thalamus, and Basal Ganglia during resting state. During symptom provocation there was a further increase of metabolism in these areas. These findings mean that in people with OCD there is increased brain disinhibition in these areas during resting state and further disinhibition during symptoms provocation. In other words, these brain regions are overactive, likely related to the symptoms of OCD. Accordingly the metabolism or brain's overactivity in the Orbito-Frontal Cortex and Caudate Nucleus of people with OCD was found to decrease after successful treatment.

Some studies have found increased metabolism in the Cerebellum as well. This is interesting since this area of the brain also serves a function of modulation, coordination or inhibition. It is possible that the activity of the Cerebellum might be increased in people with OCD as it tries to compensate for the dysfunction of the Basal Ganglia which performs similar actions. It has been proposed that people with Autism might also have a significant degree of Cerebellar dysfunction. People with Autism have characteristics in common with those who have OCD such as motor coordination deficits, obsessive thinking, and stereotyped behaviors or need to do things a certain way. Both populations seem to share Cerebellar problems. Some theorize that people with Autism share Basal Ganglia alterations in which the Basal Ganglia is trying to compensate for the Cerebellar dysfunction. However this theory has not yet been proven. By increasing understanding of these conditions society will be better able to help the people suffering from them.

There is increased Cytosolic Choline in the Medial Thalamus of people with OCD. This is a substance found inside the cells. Thus by measuring its concentration in a given region one can get an idea of the amount of cells or of their size in the tissue. Measuring for Cytosolic Choline and other chemicals in cells is how scientists are studying the minor structural differences between the brains of different people, or of the same person over time.

Ebert et al and Bartha et al have found decreased N-Acetyl Aspartate (NAA) in the Striatum (which is also part of the Basal Ganglia) of people with OCD. This substance, like the one just described, is found within cells. By measuring it scientists can have an idea of the state (size and health) of that tissue. This substance being decreased in the Striatum of people with OCD means that their Striatum either has fewer cells, or smaller ones, than people without OCD.

Using structural or anatomical studies in people with OCD researchers have found increased size of the Thalamus ("the relay station"). There is also an increase in Gray Matter in the Orbito-Frontal Cortex and the Thalamus. It is still unknown why these changes happen as they could be a result of the constant over-stimulation or part of the cause of the over-stimulation.

Using this same type of technology researchers found decreased size of the Striatum and of the Globus Pallidus (both part of the Basal Ganglia) in people with OCD. This makes sense since the Basal Ganglia in general, but more specifically the Globus Pallidus, is the part of the brain that is in charge of filtering or gating the information going to the Frontal Cortex through the Thalamus, and this seems to be impaired in people with OCD. There is also decreased Gray Matter in the Cerebellum. This is a finding that is still too preliminary to understand.

The size of the Caudate Nucleus (part of the Striatum, and more generally, the Basal Ganglia) increases after a person with OCD improves with treatment. This increase in size of the Caudate Nucleus has been found to be true with medication treatment and with psychotherapy. This observation gives more support to the theory that the Basal Ganglia is one of the primary defective brain regions in OCD. With psychosurgery, specifically with the Anterior Cingulotomy, where there is a disconnection between the Anterior Cingulate Gyrus and the Caudate Nucleus, there is a reduction in Caudate Nucleus size as the person improves. This could mean that these Genetically Inherited Procedural Behaviors or instincts are coming from the Anterior Cingulate Gyrus. By interrupting the Anterior Cingulate Gyrus' connection to the Caudate Nucleus the Caudate Nucleus decreases in size, perhaps even more than what it already was, but since there is no more stimulation from the Anterior Cingulate Gyrus, the OCD symptoms improve. Psychotherapy and medications could work in a different way. They could work by strengthening the Caudate Nucleus specifically, and Basal Ganglia in general, thus making it capable of inhibiting or controlling the stimulation by the Anterior Cingulate Gyrus. It is probable that there are many different subtypes of OCD. In some subtypes, the Basal

Ganglia is the main dysfunctional part of the circuitry which is underactive and allows a normally working Anterior Cingulate Gyrus to "leak through" stimulation. In others, the Orbito-Frontal Cortex is not sending the signal to the Basal Ganglia to stop the stimulation coming from the Anterior Cingulate Gyrus, as is the case with Orbitofrontal Meningioma. There are probably other subtypes of OCD.

Summarizing, research supports the statement that people with OCD seem to have a smaller Basal Ganglia that is abnormally active, just like the Orbito-Frontal Cortex, the Thalamus and the Anterior Cingulate Gyrus. The size of the Caudate Nucleus (part of the Striatum and hence of the Basal Ganglia) increases after successful treatment, and the level of activity decreases in all of these regions. There is also increased size of the Thalamus and increased Gray Matter in the Thalamus and in the Orbito-Frontal Cortex. There is decreased Gray Matter in the Cerebellum.

Using Transcraneal Magnetic Stimulation scientists (Greenberg et al 2000) desynchronize the brain and then measure how long it takes for it to resynchronize. They have found reduced post movement Beta synchronization. In other words, they found that in patients with OCD it took longer for the Motor Cortex (the part of the brain that has to do with initiation and maintenance of movement) to become inhibited and stable after it was destabilized or stimulated. This finding could explain why there might be difficulty in stopping an activity or movement in people with OCD, because it might take longer for their brain to go back to resting state after a behavior is started. This might explain in part why the repetition of behavior occurs.

# Pathophysiology

We propose that OCD is a set of genetically inherited motor and cognitive procedural strategies that contribute most commonly to hoarding, maintaining symmetry, checking for danger, and/or reducing contamination.

One of the functions of the Frontal-Subcortical circuits (circuits connecting the Frontal Cortex, Basal Ganglia and Thalamus) passing through the Striatum (part of the Basal Ganglia) and the Thalamus, is the execution of "prepackaged", complex, sequence-critical response behaviors that to be adaptive must be executed quickly in response to specific stimuli. These response behaviors have been named "macros". (Baxter 1992). One can experience this type of response when someone scares us. When this happens the brain immediately produces a response, even before that information reaches the Frontal Cortex and the individual is "consciously" aware of what is happening. This response is primitive and often involves a momentary paralysis. This behavior is believed to be so primitive that it is seen in a lot of insects. When some kinds of insects are threatened they become paralyzed. As a result the animal that was threatening them either loses them, or loses interest and leaves them alone. Other things that are part of that primitive and immediate response in humans are; bronchial expansion, tachycardia, hypertension, pupilary dilation, and dry mouth. The Frontal Cortex receives the information milliseconds after these responses have started and analyzes the possible threat, and if the Frontal Cortex determines that there is no threat it will inhibit all the responses that had already started. The person provoking the fight or flight response gets rewarded with the primitive reaction of the person becoming scared and is stimulated to do it again in the future. During the time that it took for the information to reach the Frontal Cortex and be analyzed, the body got itself ready for fight or flight. If the threat had been real, having this emergency response activated so quickly could have been the difference between life and death. This was especially important in the past when our ancestors lived under multiple life threatening dangers, although lamentably some people still do. Think about animal documentaries when lions ambush prey animals and these prey animals are often saved by milliseconds of reaction time.

These macros (response behaviors) require a specific pathway ("Direct Pathway") for processing this information and this pathway will take over when the appropriate stimuli are present. For example, if the roof started falling on you right now you would immediately activate your Direct Pathway System that would produce a raised heart rate and blood pressure, increased airway capacity, increased alertness, increased pupil diameter, and all of your efforts would be geared towards survival, looking for escape routes and signs of danger. Even the interesting thoughts that you are having right now would give way to the Direct Pathway System. Naturally occurring activity along the Direct Pathway would tend to rivet behavior toward the execution of the appropriate "macros" until the Frontal Cortex judged the danger (need) has passed. The Direct Pathway is usually semi-suppressed, although it is always at a resting tone or level of activity and ready to take over. It is a lot like the muscles in the body that even when not being purposely contracted, maintain a resting tone. Finding a $100 U.S. bill on the floor, in most people, would probably be an appropriate stimulus for the activation of the Direct Pathway for hoarding. If this happened to you, you would be immediately excited, more alert and awake, and your most primitive self would hope that you can keep it. At the same time this information would quickly reach the Frontal Cortex which would prompt the most appropriate response, which might not be to hoard or keep it. The Direct Pathway would be inhibited and the Indirect Pathway would be in control. This means that we function as a combination between our Direct Pathway and Indirect Pathway working at a balance. Hopefully most of us work in the Indirect Pathway most of the time but the Direct Pathway is always active at a resting tone and ready to take over or participate, if need be.

In OCD there is probably a low threshold for system "capture" by macros. There is an excess "tone" (basal level of activity) in the Direct Pathway relative to the Indirect Pathway. Macros dealing with danger, hoarding, order, and cleanliness, among others, become activated, and there is an inability to switch to other behaviors. Where these macros arise from or where they are stored in our brains, is still not evident, however, a good possible location for them might be the Anterior Cingulate Gyrus.

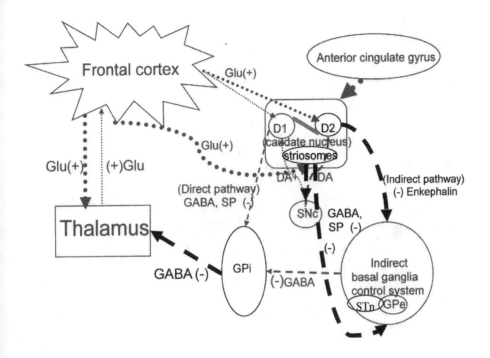

**Neurocircuitry of the Indirect Pathway. (Diagram 1)**

This is a diagram of how the brain is supposed to work most of the time. Intensity is represented by the thickness of the lines.

The Basal Ganglia is composed of the Caudate Nucleus, the Globus Pallidus Externa (GPe), the Globus Pallidus Interna (GPi), the Substantia Nigra compacta (SNc) and the Indirect Basal Ganglia Control System.

The Indirect Basal Ganglia Control System is a simplification of the connections between the Globus Pallidus Externa and the Subthalamic Nucleus (STn). They have a lot of feedback loops that are grouped together but are not clearly understood. Their main function seems to be inhibition of the Globus Pallidus Interna. The Basal Ganglia is an inhibitory, modulator system and the main function discussed in this book is the Globus Pallidus Interna inhibiting (or gating) information from going through the Thalamus to the Frontal Cortex. With the Globus Pallidus Interna exerting inhibition, the macros, instincts or procedural behaviors stay controlled. If the activity of the Globus Pallidus Interna becomes inhibited, slowed down or malfunctions for any reason, then it will not be able to perform its inhibitory function and the macros, instincts or procedural behaviors will take over.

As previously stated, there are two pathways, the Direct and the Indirect. The Indirect Pathway is the one that should be active most of the time. The main goal of the Indirect Pathway is to inhibit the actions of the Indirect Basal Ganglia Control System. In this way, the Indirect Basal

Ganglia Control System will not inhibit the actions of the Globus Pallidus Interna, and the Globus Pallidus Interna can accomplish its inhibitory function of stopping information from passing through the Thalamus to the Frontal Cortex. As seen in the diagram, there are two major projections coming from the Caudate Nucleus to the Indirect Basal Ganglia Control System. One comes from the Matrix and the other one from the Striosomes. The Matrix and the Striosomes are two different types of tissues within the Caudate Nucleus. Both of these tissues' projections are inhibitory. The one coming from the Striosomes uses the neurotransmitters GABA and Substance P (SP) to inhibit the Indirect Basal Ganglia Control System. The one coming from the Matrix uses Enkephalin to inhibit the Indirect Basal Ganglia Control System. The Dopamine 2 (D2) receptor cells in the Matrix and also the Striosomes need to be activated in order to inhibit the Indirect Basal Ganglia Control System. It is important to note that when D2 receptors are occupied by the neurotransmitter Dopamine coming from the Substantia Nigra Compacta, the effect is inhibitory on the Matrix cells. This would render the Matrix cells unable to inhibit the Indirect Basal Ganglia Control System, which in turn would increase its activity inhibiting the Globus Pallidus Interna. When Dopamine 1 (D1) receptors are occupied by the Dopamine coming from the Substantia Nigra Compacta, the effect is stimulatory on the Matrix cells, and these receptors are part of the Direct System. This means that when these receptors are stimulated or occupied by Dopamine they increase the activity of the cells in the Matrix that carry these receptors, which in turn inhibit the Globus Pallidus Interna. This allows the macros, instincts or procedural behaviors to take over. Dopamine excess tends to allow macros to take over and the person might develop OCD symptoms, or in other cases stereotypical behaviors (repetition of meaningless acts such as picking at the skin, and punding [repetitive handling and examination of objects]). The cells in the Matrix can also be stimulated by the Frontal Cortex using the neurotransmitter Glutamate. The Substantia Nigra Compacta is inhibited by the Striosomes of the Caudate Nucleus using GABA and SP. The Striosomes in turn receive Glutamate stimulation from the Frontal Cortex. In other words, the Striosomes need to be stimulated so that they inhibit the Substantia Nigra Compacta, so that the Substantia Nigra Compacta will not inhibit the Matrix and the Matrix can inhibit the Indirect Basal Ganglia

Control System. In this way the Globus Pallidus Interna will not be inhibited by the Indirect Basal Ganglia Control System and will perform its function. Also the Striosomes will directly have inhibitory capacity over the Indirect Basal Ganglia Control System.

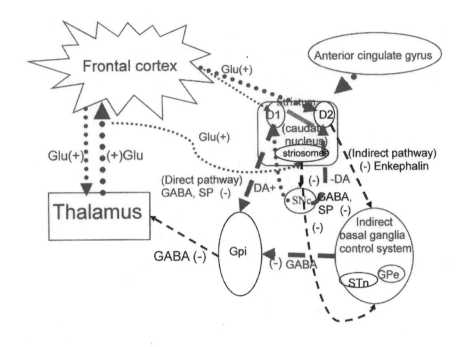

**Neurocircuitry of the Direct Pathway (Diagram 2)**

This is the same drawing. Intensity is represented by the thickness of the lines. Gray lines represent tracts that are hyperactive and black represent hypoactive tracts in OCD. The difference here is that the Striosomes activity is very deficient allowing a hyperactive Substantia Nigra to secrete Dopamine inhibiting the Matrix D2 receptors and stimulating the Matrix D1 receptors. Because the D2 receptor cells are inhibited they do not inhibit the functions of the Indirect Basal Ganglia Control System, which then inhibits the Globus Pallidus Interna. Inhibition of the Globus Pallidus Interna "opens the gate" that was keeping the macros, instincts or procedural behaviors controlled, thus allowing them to take over. Also, stimulation by the Substantia Nigra of D1 receptors cells stimulates the Matrix to inhibit the Globus Pallidus Interna through GABA and SP receptors. Thus the Globus Pallidus Interna gets inhibition from the Indirect Basal Ganglia Control System and directly from the Matrix D1 receptor cells.

Dopamine inhibits the Indirect Pathway and stimulates the Direct Pathway. When cocaine users inhale or inject cocaine they cause indiscriminate Dopaminergic stimulation. Cocaine users are actually inhibiting the Indirect Pathway and stimulating the Direct Pathway, developing in many cocaine users a sense of being in danger and stereotypical behaviors, such as picking at the skin, and punding (repetitive handling and examination of objects). Understanding the role of Dopamine in this circuitry also helps explain why antipsychotic medications are used as adjunctive or additional treatment in people with OCD. Antipsychotic medications have many actions, and one of these actions is to block the effects of the neurotransmitter Dopamine by occupying the spaces or receptors in the cells that Dopamine would have occupied. In this way these medications help the Indirect Pathway dominate over the Direct Pathway most of the time.

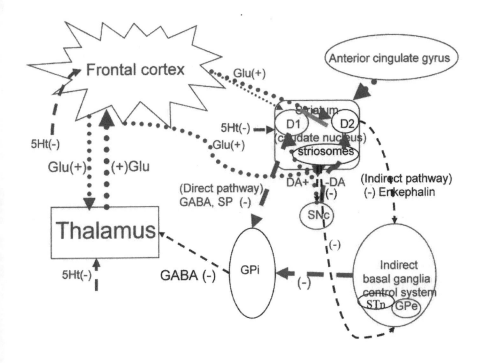

**Neurocircuitry of OCD (Diagram 3)**

This is the same neurocircuitry as the anterior one, except that this diagram includes the effect of the Serotonin Reuptake Inhibitor (5HT) medications, which are the main group of medications being used to treat OCD. This group of medications includes Fluoxetine (Prozac), Paroxetine (Paxil), Sertraline (Zoloft), Citalopram (Celexa), Estacitalopram

(Lexapro), Fluvoxamine (Luvox), and Vilazodone (VIIBRYD). New medications are constantly being developed. There is an older medication named Clomipramine which is effective in the treatment of OCD but due to a higher incidence and severity of side effects, the use of this medicine has been declining over the years. Other medications that are some times used are the combined norepinephrine/serotonin reuptake inhibitors such as Venlafaxine (Effexor) and Desvenlafaxine (Pristiq). There are a lot of serotonergic cells sending their axons towards the Frontal Cortex, the Thalamus and the Striatum. Serotonin is modulating/inhibitory thus possibly decreasing the circuitry's over-stimulation. Serotonin also modulates or inhibits dopaminergic cells such as the ones coming from the Substantia Nigra Compacta, and thus decreases the effect of the Substantia Nigra Compacta and favors functioning of the Indirect Pathway.

One can notice the weak Globus Pallidus Interna inhibitory function over the Thalamus, which is probably the main problem in OCD. Inhibition of the Globus Pallidus Interna as stated before, removes the gate that keeps the macros, instincts or procedural behaviors controlled, which allows them to take over. In OCD this inhibition of the Globus Pallidus Interna, or the inability of the Globus Pallidus Interna to function appropriately is prolonged, causing the macros, instincts or procedural behaviors to take over for a longer time, in a maladaptive way. In other words they are active in situations when they should not be active and for a longer period of time. This diagram also shows the strong inhibition by the Indirect Basal Ganglia Control System and the Dopamine 1 receptor cells from the Matrix of the Caudate Nucleus, over the Globus Pallidus Interna. This is when we see the "Brain Lock" described by Jeffrey M. Schwartz, M.D. in his book BRAIN LOCK. In his book, Dr. Schwartz makes an analogy between the malfunctioning brain in OCD and a brain that is locked, or unable to shift to healthier functioning.

Since the brain is a closed circuit, any dysfunctional part in the circuit will have an effect on the rest of it. One can see from the neurocircuitry diagrams how failure of any of the following could cause OCD:

Failure of the Orbito-Frontal Cortex to stimulate the Caudate Nucleus, as in Orbito-Frontal Meningioma, will cause an increase in Dopamine activity leading to weak function of the Globus Pallidus Interna.

Failure of the Striosomes, or of the D2 receptor cells to respond, as might happen in Huntington's disease, will also cause a weak functioning of the Globus Pallidus Interna.

Failure of the Globus Pallidus Interna itself, as in bilateral necrosis of the Globus Pallidus, will allow macros to go to the frontal cortex disinhibited.

Too much Dopamine, as in Cocaine users, will also cause inhibition of the Globus Pallidus Interna.

Failure of the overall inhibitory function of the Basal Ganglia, as might happen in PANDAS, and more specifically in Sydenham's chorea, allows passage of disinhibited macros to the Frontal Cortex.

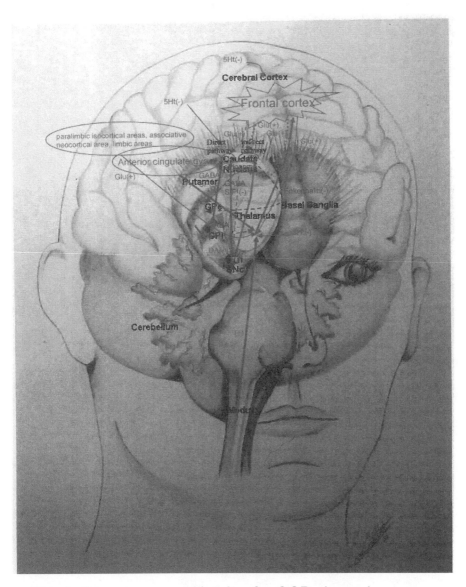

This is the same neurocircuitry for OCD shown in a more anatomically correct way

# Treatment

Previously in this book the brain was compared to a computer. One of the great differences between the two is that the brain is a biological entity, a living thing, and it physically changes to a certain extent over time. The brain changes its "wiring" due to multiple factors such as experiences, learning, training, illnesses, drugs, medications, trauma, and plain old time. Because of this ability of the brain to change, it is very important to care for it and respect it, since the brain could change in a good or in a bad way. Wise people use this brain quality to improve their own brains, as well as the brains of people with whom the world is shared.

Psychotherapy is a form of treatment that involves thinking, talking, and actions, to improve a person's condition, or ability to function in life. Studies have proven that psychotherapy can actually change, to a limited extent, the chemistry and the structure of the brain, and this could make a huge difference in someone's life. Although some people believe that this is an innocuous form of treatment, it is not. Several types of psychotherapies may be used with any given person. If the type of therapy is not chosen and applied correctly, it could contribute to a worsening in the person's condition or the triggering of other conditions. It could also lead to increased shame, guilt, helplessness, and hopelessness. Selecting the appropriate form of therapy requires a partnership between therapist and patient in which both parties take into consideration the effective forms of therapy, patient's preference, and therapist proficiency.

The form of psychotherapy that has proven most efficacious for OCD in more scientific studies is Cognitive Behavioral Therapy (CBT). This treatment can be quite structured or manualized. It involves education about what the illness is and what it is not. Cognitive Behavioral Therapy educates the person about the biological condition, or illness that has taken over his/her brain and life. It is important for the patient to understand that an area of the brain is inappropriately taking over control, and needs to be subdued through treatment. The person should think of OCD as an intruder in the brain that must be defeated. The psychotherapist is a tool to conquer the illness, but the ultimate responsibility in defeating OCD lies with the patient.

The person with OCD has different stimuli that trigger the compulsions and obsessions. These stimuli are identified and graded in severity. The person and the therapist identify how strongly the stimuli stimulate the person to perform the compulsions and elicit the obsessions, or how difficult it is for the person to resist them.

There is a technique called Exposure and Relapse Prevention (ERP). In this technique the person is exposed to the feared, or response provoking stimulus, with the goal that he/she has to refrain from performing the compulsion. For example if a person's compulsion is about cleanliness, he or she is asked to touch the floor and refrain from cleaning the hands until the desire and anxiety to clean them has subsided. This is a type of training that starts with the easiest behavior, and once that behavior is controlled, then one progresses to harder ones. After an exposure to a feared stimulus, anxiety will be very high. During this time the person is expected to apply the knowledge gained about the illness. The person understands that these impulses and ideas are inappropriate and although they feel real, the person knows that they are not. The goal, as mentioned above, is to refrain from performing the compulsion until the anxiety decreases and finally subsides. Many patients find it helpful to engage themselves in activities that will capture their attention, taking their attention away from OCD related impulses and ideas. Some of these activities may be talking with someone else such as family members or a friend, getting physically away from the possibility of performing the compulsion, and doing physical activities, to name a few. One can get on a bicycle and ride down hill. By the time one makes it back up the hill, the compulsion hopefully will have subsided. Ignoring the impulse is extremely hard to do, but if the person gives in and performs the compulsion, the faulty brain mechanisms responsible for this continued behavior will strengthen and the illness only worsens. On the other hand, if the intense impulse is endured and the compulsion is not performed, the brain's inhibitory mechanisms are strengthened making the compulsion and obsession weaker and easier to tolerate. Frequently, even after successful treatment, the impulse to perform the compulsions, and the accompanying obsessions, does not go away completely but it is weaker and easier to control, which allows the person to live a more "normal life".

Cognitive Behavioral Therapy could work by changing the significance, meaning or importance of the stimulation being received by the Frontal Cortex, even if the Frontal Cortex continues to receive the stimulation. The Frontal Cortex could also change brain activity in other areas, such as inhibiting the Hypothalamus from activating the Autonomic Nervous System, and decreasing the activity of the Amygdala. Also, by stopping a procedural behavior or altering it, its procedural nature stops being reinforced and inhibiting mechanisms are strengthened, thus actually decreasing the stimulation being received by the Frontal Cortex

For more on the psychotherapy of OCD refer to BRAIN LOCK by Jeffrey M. Schwartz, M.D. with Beverly Beyette, and OCD IN CHILDREN AND ADOLESCENTS: A COGNITIVE-BEHAVIORAL TREATMENT MANUAL by John S. March, Karen Mulle. After discontinuation of Cognitive Behavioral Therapy treatment the relapse rate is less than in the people treated with medications alone.

Medications also have a role in the treatment of OCD, especially the newer antidepressants named as a group

Serotonin Reuptake Inhibitors (SRI). Their name indicates their major mode of action in the body. Serotonin is a neurotransmitter, one of the substances that communicate one nerve cell with the next. This chemical is removed from its site of action after its function is performed and recovered by the cell that produced it. The SRI medications interrupt this reuptake allowing for more Serotonin to remain in the site of action. What happens from here is not completely understood but here are some of the theories. The Serotonin System has modulating/inhibitory effects and has multiple and important connections to the Thalamus, Cortex and Caudate Nucleus, thus decreasing the system's over-stimulation. The Serotonin system also inhibits dopaminergic cells, thus decreasing the activity in the dopaminergic system as well. Serotonin stimulates cells to produce Neurotropic Derived Growth Factor, causing neurons to produce more dendrites, which may help explain the findings of an increase in size of the Caudate Nucleus. Dendrites are the part of the neurons that grow opposite to the axons. If compared to a tree, the Dendrites would be the branches and the axons its trunk and roots. Usually medications have to be taken for many months and often years. These medications are quite safe when compared to other medications such as antibiotics or cardiovascular medications, but they should always be prescribed and monitored by a physician, preferably a psychiatrist. Remember that these medicines reach the brain which is one of the best protected organs of the body. If they get into the brain, they will get virtually everywhere in the body, including body fluids, such as milk during lactation.

Medications, when used, should ideally be in conjunction with Cognitive Behavioral Therapy. People who undergo a combined treatment may require fewer medications and experience a longer time to relapse after stopping treatment. At times medications are used as the sole form of therapy in people who for many different reasons are not able to have the recommended Cognitive Behavioral Therapy. Medications alone could be effective but as stated before, the relapse rate after stopping them is higher than when they are combined with Cognitive Behavioral Therapy. Medications are often used with people whose illness is so severe that they cannot successfully apply the techniques of Cognitive Behavioral Therapy. These substances can decrease the anxiety and the level of inappropriate stimulation to a state where the person can effectively do Cognitive Behavioral Therapy. They are usually used at dosages much higher than those used to treat depression and the illness takes longer to improve, maybe due to the fact that the medication is increased gradually until a safe and effective dose is reached. Once treatment has begun the person usually feels nothing. Usually the good or therapeutic effects do not start until up to at least one to two weeks of treatment, but usually longer. These medicines (Serotonin Reuptake Inhibitors) do not work on an as needed basis and need to be taken everyday for them to work, otherwise the money and the effort are being wasted. Some times an individual may feel mild side effects (unwanted effects) of the medicine. Most of the side effects, when experienced, usually happen during the first weeks of treatment and then improve or totally go away. Side effects can also happen when the dose of the medication is increased. Side effects are dose related, meaning that the higher the dose the more likely it is that a person will experience them. Physicians should generally start the medication at a low dose and increase it slowly as tolerated to decrease the risk of side effects. Some of the most common side effects include worsening of anxiety, abdominal pain, diarrhea, dry mouth, and sexual dysfunction. Most people do not develop these side effects and the ones that do experience them, the side effects are usually mild. There are a few life threatening reactions that fortunately are very uncommon. Among them are the Serotonin Syndrome and the Neuroleptic Malignant Sydrome (NMS), which are both life threatening and hence medical emergencies.

Serotonin Syndrome is Serotonin excess or intoxication. This syndrome can happen with multiple medications (some not typically considered as psychotropic medications) and even with street drugs. People that develop Serotonin Syndrome can experience increased heart rate, myoclonus (body jerks), sweating, dilated pupils, over responsive reflexes, hyperactive bowels, high blood pressure, hypervigilance, agitation, and severely high body temperatures. If the temperature is not lowered successfully, the person can develop muscle breakdown, seizures, renal failure, disseminated intravascular coagulation, and eventually die. The first step in treating Serotonin Syndrome is discontinuation of the substance that caused it, and usually the person needs hospitalization.

Neuroleptic Malignant Syndrome is usually caused by antipsychotic medications, but can also be caused by antidepressants, by medications not considered psychotropic medications, and even by street drugs. Symptoms include fever, autonomic instability, and cognitive changes such as delirium. There is usually elevation of plasma creatine phosphokinase and white blood cell count. There is also possibility of complications as seen in Serotonin Syndrome due to high body temperature and muscle breakdown. Neuroleptic Malignant Syndrome also is a medical emergency as it can be fatal. It will most likely require hospitalization and intensive medical treatment.

There are other uncommon reactions that are beyond the scope of this book. A detailed description of side effects and reactions should be discussed with your physician.

In people with a severe and refractory condition (did not respond to conventional medication treatment), other medications might need to be used. Among them are the Antipsychotics. These are medications that among other actions, block the effects of Dopamine. Although this is not the typical use for these medications they can be used in conjunction with the SRI to increase their efficacy. Probably medications with strong blockage of Dopamine receptors will be preferable, such as Risperidone (Risperdal). They are usually used at low dosages. Dopamine blockage would decrease the symptoms of OCD by decreasing the effect of the Substantia Nigra's enervation on the caudate nucleus, thus facilitating Indirect Pathway activity and inhibition of the Direct Pathway.

Psychosurgery, up to this date is rarely used and only in cases that are severe and refractory, or unresponsive to all other forms of treatment. It is considered the most invasive form of treatment. In this intervention the neurosurgeon disrupts the activity of the neurons transmitting impulses or information among the different brain regions. How this disruption is achieved has been evolving over time as previously discussed. Today the rate of intra and post surgical complications has markedly decreased making these surgeries safer.

Other treatments that are tried by clinicians and researchers are Electro Convulsive Therapy (ECT), relaxation techniques, hypnosis, and family therapy with some success.

As with many medical therapies, none of the treatments mentioned in this book are one hundred percent effective for everyone, not even in combination.

# Conclusion

Individuals who have OCD or who know someone with it are not alone, even though it might feel that way at times. OCD is an illness that affects 2 to 4 percent of the adult population in the USA, or approximately 4 million adult Americans. In addition it is estimated that there are 1.3 million children. This makes it more common than Schizophrenia and Bipolar Disorder combined. It equally affects men and women of all socioeconomic groups. It is the 9$^{th}$ leading cause of disability in the USA and 10$^{th}$ in the world. According to many studies, between 5 to 25 percent of people with OCD attempt suicide. These statistics only speak of the person with OCD and do not include the ripple effects that the illness has on family members and society at large.

OCD is not a character or personality weakness. It is a medical disorder that continues to be extensively researched. It is not, as once believed, caused by childhood experiences, although people with OCD commonly experience some aspects of the illness in their childhood because it can be genetically inherited. Their parents may have had the illness themselves or displayed some traits of it. OCD is not only genetic, it can surface secondary to another condition such as an infection, tumor, strokes, and another genetic disorder such as Huntington's disease, to name a few.

Failure in any of multiple components of the Orbito-Frontal Cortex Subcortical circuit can cause symptoms of OCD as is seen in the diagrams and also from the medical disorders associated with OCD. Multiple abnormalities can cause OCD, which means that there are probably multiple neuroanatomic and neuropathologic subtypes of OCD.

Cognitive Behavioral Therapy is well established as effective therapy for people with OCD. Medication treatment is also effective and when used, it should ideally be done in combination with Cognitive Behavioral Therapy. The initial medications to be used in most people should be the SRIs, and if these do not work, additional medications might need to be used such as the Antipsychotics, and at times the psychostimulants. In people with severe illness who are refractory to every other treatment, psychosurgery may be an option.

This neurobiological model of OCD is proposed as a possible way of thinking about the condition based on current scientific literature. The hope is that this book will help the people suffering from this condition, their loved ones, as well as stimulate thinking, discussion and further research among scientists and clinicians.

## Future Research

In recent studies the Cerebellum has shown abnormalities in some patients with OCD. The Cerebellum is also being identified as being involved in the pathology of Autism which shares some characteristics with OCD. More research needs to be done about the higher functions of the Cerebellum and the Basal Ganglia. The relationship between OCD and coordination deficits as well as between Autism and OCD like symptoms (such as perseveration and cognitive rigidity) needs to be further studied.

Neuroimaging and neurosurgical studies will eventually give more specific information about the locations of the damaged brain structures, helping scientists classify the different subtypes of OCD. This will allow more targeted treatments for individuals.

New medications need to be developed for the different OCD subtypes that will exert their effect in more specific brain pathways. This will increase their effectiveness, aid in the avoidance of uncomfortable side effects, and increase their safety.

More studies need to be performed to study the changes that Cognitive Behavioral Therapy causes in the brain after successful treatment. Already there is data indicating that Cognitive Behavioral Therapy causes measurable changes in the caudate nucleus, but it could do it also in the Frontal Cortex, the Anterior Cingulate Gyrus, and/or other areas.

In the future, scientists will probably be able to map where these critical response primitive behaviors (genetically inherited procedural behaviors or instincts) are stored in the brain, and will probably be able to stimulate them in anyone by stimulation of the appropriate brain region. This is already being done with predatory behaviors in cats, where Greg TR, Siegel A. (2001) has implanted electrodes in a cat's brain and has been able to stimulate the electrode, causing the cat to perform the act of grabbing a mouse by the neck in an automatic fashion.

# Bibliography

1- Abbot, A.: Brain implants show promise against obsessive disorder. Nature. 419:658, 2002.

2- Alcock, J. (1998) *Animal Behavior: An Evolutionary Approach* (6th edition), Chapter 5. Sinauer Associates, Inc. Sunderland, Massachusetts.

3- Anxiety Disorders Association of America http://www.adaa.org/mediaroom/index.cfm

4- Bartha R, Stein MB, Williamson PC, et all: A short echo 1H spectroscopy and volumetric MRI study of the corpus striatum in patients with obsessive-compulsive disorder and comparison subjects. Am J Psychiatry 155:1584-1591, 1998.

5- Baxter LR, Schwartz JM, Guze BH, et al: Caudate glucose metabolic rate canges with both drug and behavior therapy for obsessive-compulsive disorder. Arch Gen Psychiatry 49:681-689, 1992.

6- Belluardo N., Mudo G., Blum M., Amato G., Fuxe K.: Neurotrophic effects of central nicotine receptor activation. Journal of Neural Transmission 60:227-45, 2000. Brown, S., VMD; Crowell-Davis, S., DMV, PhD; Malcolm, T., DVM: Naloxone-responsive compulsive tail cashing in a dog. JAVMA, Vol 190, No. 7, April 1, 1987.

7- Benkelfat C, Murphy DL, Zohar J, Hill JL, Grover G, Insel. TR.Clomipramine in obsessive-compulsive disorder. Further evidence for a serotonergic mechanism of action. Arch Gen Psychiatry. Jan; 46(1):23-8, 1989.

8- Busatto, G., Ph.D. Buchpiguel, C, Ph.D, Zamignani D., B.A, et. al.: Regional cerebral blood flow abnormalities in early-onset obsessive-compulsive disorder: An exploratory SPECT study. Journal of the American Academy of Child and Adolescent Psychiatry, 40(3):347-354, 2001.

9- Carmichael S., Price J.: Limbic connections of the orbital and medial prefrontal cortex in Macaque Monkeys. The Journal of Comparative Neurology, 363:615-641, 1995.

10- Cosgrove G., MD, Rauch S., MD: Psychosurgery. Departments of Neurosurgery and Psychiatry, Massachusetts General Hospital and Harvard Medical School.

11- Dallaire, A.: Stress and behavior in domestic animals. Temperament as a predisposing factor to stereotypies. Annals New York Academy of Sciences. 269-274.

12- Dinn. W., Harris C., Raynard R.: Posttraumatic obsessive-compulsive disorder: A three-factor model. Psychiatry, 62:313-321, 1999.

13- Dodman, N. H., Moon-Fanelli, A., and Mertens, P. A., Pflueger, S., Stein, D. J.: Veterinary Models of OCD. Veterinary Models pp99-143.

14- Ebert D., Speck O., König A., Berger M., et al: 1H-magnetic resonance spectroscopy in obsessive-compulsive disorder: evidence for neuronal loss in the cingulate gyrus and the right striatum. Psychiatry Res. 1997 Jul 4;74(3):173-6.

15- Francobandiea, MD., Quetiapine augmentation of sertraline in obsessive-compulsive disorder (Letter to the editor). Journal of Clinical Psychiatry 63(11)1046-47, 2002.

16- Gadow K., Ph.D., Nolan E., Ph.D., Sprafkin J., Ph.D., Schwartz J., Ph.D.: Tics and psychiatric co morbidity in children and adolescents. Developmental Medicine & Child Neurology 44:330-338, 2002.

17- Gilbert A., MD, Moore G., PhD, Keshavan M., MD: Decrease in thalamic volumes of pediatric patients with obsessive-compulsive disorder who are taking Paroxetine. Arch Gen Psychiatry 57:449-457, 2000.

18- Goldman-Rakic PS. Circuitry of primate prefrontal cortex and regulation of behavior by representational memory. In: Handbook of Physiology. A Critical Comprehensive Presentation of Physiological Knowledge and Concepts. Section 1: The Nervous System, vol. V. Higher Functions of the Brain, Part I, American Physiological Society, Bethesda, 1987

19- Goldberger E, Rapoport JL: Canine Acral Lick Dermatitis: Response to anti-obsessional drug Clomipramine. J Am Anim Hosp Assoc, 1990, in press

20- Greenberg B. MD, PhD, Ziemann U.: Decreased neuronal inhibition in cerebral cortex in obsessive-compulsive disorder on transcranial magnetic stimulation. The Lancet 352:881-882, 1998.

21- Greenberg B., MD, PhD, Murphy D., MD: Neuroanatomically based approaches to obsessive-compulsive disorder. The Psychiatric Clinics of North America 23:671-685, 2000.

22- Greenberg B., MD, PhD, Ziemann U., MD: Altered cortical excitability in obsessive-compulsive disorder. American Academy of Neurology 54:142-148, 2000.

23- Greg TR, Siegel A.: Differential effects of NK1 receptors in the midbrain periaqueductal gray upon defensive rage and predatory attack in the cat. Brain Res. 2003 Dec 19;994(1):55-66.

24- Greg TR, Siegel A.: Brain structures and neurotransmitters regulating aggression in cats: implications for human aggression. Prog Neuropsychopharmacol Biol Psychiatry. 2001 Jan;25(1):91-140.

25- Insel, T. M.D.; Mos, J.; Olivier, B.: Animal Models of Obsessive Compulsive Disorder: A Review. Pathophysiology. 1994.

26- Insel, T., M.D.: Obsessive-Compulsive Disorder: New Models. Psychopharmacology Bulletin Vol. 24, No. 3, 1988: pp 365-369

27- Insel T., MD: Toward a neuroanatomy of obsessive-compulsive disorder. Arch Gen Psychiatry 49:739-746, 1992.

28- Joseph R.: Frontal lobe psychopathology: Mania, Depression, Confabulation, Catatonia, Perseveration, Obsessive Compulsions, and Schizophrenia. Psychiatry, 62:138-172, 1999.

29- Kaplan & Sadock's Comprehensive Textbook of Psychiatry, seventh ed. Philadelphia: Lippincott Williams & Wilkins, 2000.

30- Khullar A., MD, Chue P., MB Bch, Tibbo P., MD: Quetiapine and obsessive-compulsive symptoms (OCS): case report and review of atypical antipsychotic-induced OCS. Journal of Psychiatry & Neuroscience, 8(1):55-59, 2001.

31- Kim J., Myung Chul L., Jaeseok K., et. al: Grey matter abnormalities in obsessive-compulsive disorder: Statistical parametric mapping of segmented magnetic resonance images. The Royal of College of Psychiatrists, 179:330-334, 2001.

32- Konig A., Thiel T., Ebert D., Overmeier S.: Magnetic resonance spectroscopy of the right striatum in obsessive-compulsive disorder: the role of the Basal Ganglia. German Journal of Psychiatry 1(2):53-61, 1998.

33- Leocani, L. M.D., PhD.; Locatelli, M. M.D.; Bellodi, L. M.D.; et al: Abnormal pattern of cortical activation associated with voluntary movement in Obsessive-compulsive disorder: An EEG study. Am. J. of Psychiatry, Volume 158(1). January 2001.140-142.

34- Luescher, U. D.V.M., PhD.; Mckeown, D. D.V.M.; Halip, J. D.D.S., MScD.: Stereotypic or Obsessive-Compulsive Disorders in dogs and cats. Veterinary

Clinics of North America: Small Animal Practice. Vol. 21, No. 2, March 1991.

35- McGrath M., Campbell K., Parks C., Burton F.: Glutamatergic drugs exacerbate symptomatic behavior in a transgenic model of co morbid Tourette's syndrome and obsessive-compulsive disorder. Brain Research, 877(1):23-30, 2000.

36- Montoya A., MD., Weiss A., MD., Price, B., MD, et al: Magnetic Resonance imaging-guided stereotactic limbic leucotomy for treatment of intractable psychiatric disease. Neurosurgery, 50(5) 1043-1052, 2002.

37- NAMI
http://www.nami.org/Content/NavigationMenu/Inform_Yourself/About_Public_Policy/Policy_Research_Institute/Policymakers_Toolkit/Facts_for_Policymakers_Treatable_Causes_of_Disability_-_Anxiety_Disorders.htm

38- Nieuwenhuys R., MD, PhD, Voogd J., MD, PhD: The Human Central Nervous System. Third revised edition. Springer-Verlag Berlin Heidelberg, Germany, pp 247-258

39- Noback C., Strominger N., Demarest R.: The Human Nervous System. Introduction and Review. Fourth Edition. LEA & Febiger, Philadelphia, pp 191, 1991.

40- Oades R., Ropcke B., Eggers C.: Monoamine activity reflected in urine of young patients with obsessive-compulsive disorder, psychosis with and without reality distortion and healthy subjects: An explorative analysis. Journal of Neural Transmission 96:143-149, 1994.

41- Panel Reports: Lick granuloma. Modern Veterinary Practice 55:139-145, 1974

42- Pato, M., M.D.; Pato, C., M.D.; Pauls, D., PhD.: Recent findings in the genetics of OCD. J. Clin. Psychiatry 2002:63 (suppl 6).

43- Rauch S., MD, Kim H. MA, Makris N., MD, Ph.D. et al: Volume reduction in the caudate nucleus following stereotactic placement of lesions in the Anterior Cingulate Cortex in humans: a morphometric magnetic resonance imaging study. Neurosurgery, 93:1019-25, 2000.

44-Rapoport, J. M.D.: Recent advances in Obsessive-Compulsive Disorder. Neuropsychopharmacology 1991-Vol/ 5, No.1

45-Rapoport, J. M.D.; Insel, T. M.D.: Obsessive-Compulsive Disorder: New models. Obsessive-Compulsive Disorder: A neuroethological perspective. Psychopharmacology Bulletin. Vol. 24, No. 3, 1988.

46-Ressler K., Nemeroff C.: Role of serotonergic and noradrenergic systems in the pathophysiology of depression and anxiety disorders. Depression & Anxiety. 12(1):2-19, 2000.

47-Riad M. Emerit MB, Hamon M.: Neurotropic effects of Ipsapirone and other 5-HT1A receptor agonists on septal cholinergic neurons in culture. Brain Research. Developmental Brain Research. 82(1-2):245-258, 1994.

48-Ridley, R., Baker, H.: Stereotypy in monkeys and humans. Psychological medicine, 1982, 12, 61-72

49-Robinson D., MD, Wu H,., MD, Munne R., MD: Reduced caudate nucleus volume in obsessive-compulsive disorder. Arch Gen Psychiatry 52:393-398, 1995.

50- Rosenberg D., Keshavan M., Dick E., Bagwell W., MacMaster F., Birmaher B., Corpus Callosum morphology in treatment-naïve pediatric obsessive-compulsive disorder. Prog. Neuro-Psychophamacology & Biological Psychiatry, 21:1269-83, 1997.

51- Rosenberg D., MacMaster R., Deshavan M., Fitzgerald K., Moore S.: Decrease in Caudate Glutamatergic concentrations in pediatric obsessive-compulsive disorder in patients taking Paroxetine. Journal of American Academy of Child and Adolescent Psychiatry, 39(9):1096-103, 2000.

52- Rosenberg, D, MD, Amponsah A., BS, Sullivan A., BA, et al: Increased Medial Thalamic Choline in pediatric obsessive-compulsive disorder as detected by quantitative in vivo spectroscopic imaging. Journal of Child Neurology 16(9) 636-641, 2001.

53-Saxena S., MD, Rauch S., MD: Functional neuroimaging and neuroanatomy of obsessive-compulsive disorders. The Psychiatric Clinics of North America 23:563-584, 2000.

54-Sharma, P.; Gupta, N.: Obsessive-compulsive phenomenon and Parkinson's disease. J. Neurol Neurosurg Psychiatry, Volume 72(3). March 2002.420.

55-Shidara, M.; Richmond, B.: Anterior Cingulate: Single neuronal signals related to degree of reward expectancy. Science, Volume 296(5573). May 31, 2002. 1709-1711.

56-Shaikh MB, Siegel, A.: Neuroanatomical and neurochemical mechanisms underlying amygdaloid control of defensive rage behavior in the cat. Brazilian journal of Medical & Biological Research. 27(12):2759-79, 1994.

57-Stein, D., M.D.; Dodman, N.; Borchelt, P.; et al.: Behavioral disorders in veterinary practice: Relevance to Psychiatry. Comprehensive Psychiatry, Vol. 35. No. 4 (July/August), 1994: pp 275-285.

58- Stein, D., MD, Ludik, J.: A neural network of obsessive-compulsive disorder: Modelling cognitive disinhibition and neurotransmitter dysfunction. Medical Hypothesis 55(2), 168-176, 2000.

59- Stein, D., MD: ADVANCES IN THE NEUROBIOLGY OF OBSESSIVE-COMPULSIVE DISORDER. The Psychiatric Clinics of North America 23:545-562, 2000.

60- Stein, D., M.D.: Obsessive-compulsive disorder. The Lancet, 360:397-404, 2002.

61- Stein, D., M.D.; Shoulberg, N.; Helton, K.; et al.: The neuroethological approach to Obsessive-Compulsive Disorder. Comprehensive Psychiatry. Vol. 33, No. 4 (July/August), 1992: pp 274-281.

62- Swedo S, Rapoport J, Leonard H, Lenane M, Cheslow D, (1989a): Obsessive compulsive disorder in children and adolescents: Clinical phenomenology of 70 consecutive cases. Arch Gen Psychiatry 46:335-341

63-Szeszko, P., PhD, Robinson, D., MD, Alvir, J, DrPH: Orbital Frontal and Amygdala volume reductions in obsesssive-compulsive disorder. Arch Gen Psychiatry 56:913-920, 1999.

64-Tinbergen, N. (1951) *The Study of Instinct*. Oxford University Press, New York.

Made in the USA
Middletown, DE
20 October 2024